AF876039

Cell Death and Disease

Edited by Ke Xu

Published in London, United Kingdom

Cell Death and Disease
http://dx.doi.org/10.5772/intechopen.104360
Edited by Ke Xu

Contributors
Leisheng Zhang, Baoquan Song, Yanbo Shi, Junyong Zhang, Kaitao Lu, Sunfeng Pan, Hanqiang Shi, Lie Xiong, Shuqin Du, Ting Li, Qirong Wen, Qingfa Zeng, Chayan Munshi, Shelley Bhattacharya, Vincenzo Carafa, Lucia Altucci, Antonio Beato, Gregorio Favale, Giulia Verrilli, Laura Della Torre, Vincenza Capone, Daniela Carannante, Silva Samson Abrahamyan, Karina Galoian

© The Editor(s) and the Author(s) 2023
The rights of the editor(s) and the author(s) have been asserted in accordance with the Copyright, Designs and Patents Act 1988. All rights to the book as a whole are reserved by INTECHOPEN LIMITED. The book as a whole (compilation) cannot be reproduced, distributed or used for commercial or non-commercial purposes without INTECHOPEN LIMITED's written permission. Enquiries concerning the use of the book should be directed to INTECHOPEN LIMITED rights and permissions department (permissions@intechopen.com).
Violations are liable to prosecution under the governing Copyright Law.

Individual chapters of this publication are distributed under the terms of the Creative Commons Attribution 3.0 Unported License which permits commercial use, distribution and reproduction of the individual chapters, provided the original author(s) and source publication are appropriately acknowledged. If so indicated, certain images may not be included under the Creative Commons license. In such cases users will need to obtain permission from the license holder to reproduce the material. More details and guidelines concerning content reuse and adaptation can be found at http://www.intechopen.com/copyright-policy.html.

Notice
Statements and opinions expressed in the chapters are these of the individual contributors and not necessarily those of the editors or publisher. No responsibility is accepted for the accuracy of information contained in the published chapters. The publisher assumes no responsibility for any damage or injury to persons or property arising out of the use of any materials, instructions, methods or ideas contained in the book.

First published in London, United Kingdom, 2023 by IntechOpen
IntechOpen is the global imprint of INTECHOPEN LIMITED, registered in England and Wales, registration number: 11086078, 5 Princes Gate Court, London, SW7 2QJ, United Kingdom

British Library Cataloguing-in-Publication Data
A catalogue record for this book is available from the British Library

Additional hard and PDF copies can be obtained from orders@intechopen.com

Cell Death and Disease
Edited by Ke Xu
p. cm.

This title is part of the Biochemistry Book Series, Volume 39
Topic: Cell and Molecular Biology
Series Editor: Miroslav Blumenberg
Topic Editor: Rosa María Martínez-Espinosa

Print ISBN 978-1-83768-244-7
Online ISBN 978-1-83768-245-4
eBook (PDF) ISBN 978-1-83768-246-1
ISSN 2632-0983

We are IntechOpen, the world's leading publisher of Open Access books
Built by scientists, for scientists

6,200+
Open access books available

169,000+
International authors and editors

185M+
Downloads

156
Countries delivered to

Our authors are among the
Top 1%
most cited scientists

12.2%
Contributors from top 500 universities

WEB OF SCIENCE™

Selection of our books indexed in the Book Citation Index
in Web of Science™ Core Collection (BKCI)

Interested in publishing with us?
Contact book.department@intechopen.com

Numbers displayed above are based on latest data collected.
For more information visit www.intechopen.com

IntechOpen Book Series

Biochemistry

Volume 39

Aims and Scope of the Series

Biochemistry, the study of chemical transformations occurring within living organisms, impacts all of the life sciences, from molecular crystallography and genetics, to ecology, medicine and population biology. Biochemistry studies macromolecules - proteins, nucleic acids, carbohydrates and lipids –their building blocks, structures, functions and interactions. Much of biochemistry is devoted to enzymes, proteins that catalyze chemical reactions, enzyme structures, mechanisms of action and their roles within cells. Biochemistry also studies small signaling molecules, coenzymes, inhibitors, vitamins and hormones, which play roles in the life process. Biochemical experimentation, besides coopting the methods of classical chemistry, e.g., chromatography, adopted new techniques, e.g., X-ray diffraction, electron microscopy, NMR, radioisotopes, and developed sophisticated microbial genetic tools, e.g., auxotroph mutants and their revertants, fermentation, etc. More recently, biochemistry embraced the 'big data' omics systems. Initial biochemical studies have been exclusively analytic: dissecting, purifying and examining individual components of a biological system; in exemplary words of Efraim Racker, (1913 –1991) "Don't waste clean thinking on dirty enzymes." Today, however, biochemistry is becoming more agglomerative and comprehensive, setting out to integrate and describe fully a particular biological system. The 'big data' metabolomics can define the complement of small molecules, e.g., in a soil or biofilm sample; proteomics can distinguish all the proteins comprising e.g., serum; metagenomics can identify all the genes in a complex environment e.g., the bovine rumen.
This Biochemistry Series will address both the current research on biomolecules, and the emerging trends with great promise.

Meet the Series Editor

Miroslav Blumenberg, Ph.D., was born in Subotica and received his BSc in Belgrade, Yugoslavia. He completed his Ph.D. at MIT in Organic Chemistry; he followed up his Ph.D. with two postdoctoral study periods at Stanford University. Since 1983, he has been a faculty member of the RO Perelman Department of Dermatology, NYU School of Medicine, where he is codirector of a training grant in cutaneous biology. Dr. Blumenberg's research is focused on the epidermis, expression of keratin genes, transcription profiling, keratinocyte differentiation, inflammatory diseases and cancers, and most recently the effects of the microbiome on the skin. He has published more than 100 peer-reviewed research articles and graduated numerous Ph.D. and postdoctoral students.

Meet the Volume Editor

Professor Ke Xu earned his BSc in Microbiology from Nankai University, China, and his Ph.D. in Cell and Molecular Biology from the University of Essex, UK. He completed his postdoctoral training at the Institute of Cancer Research, UK, working on leukemia. Professor Xu carried out his research fellowship at Imperial College London, investigating gene targeting and lung cancer. He joined the Tianjin Lung Cancer Institute of Tianjin Medical University General Hospital, China in 2007 as a principal investigator, studying tumor microenvironment and novel anti-cancer drugs. Professor Xu is an active member of the American Association for Cancer Research, European Association for Cancer Research, American Society for Cell Biology, and Chinese Anti-Cancer Association.

Contents

Preface

Cell death plays an important role in various physiological and pathological conditions. It is a process involving the growth and differentiation of cells, death of normal cells, and cleaning of abnormal cells. The dysregulation of the cell death process may lead to diverse diseases.

There are two main types of cell death: necrosis and programmed cell death (PCD). Necrosis is an unregulated, passive process characterized by organelle and cell membrane destruction. PCD is an active extinction process, including apoptosis, autophagy, ferroptosis, pyroptosis, and necroptosis. Among them, the most studied PCD processes are apoptosis, autophagy, and ferroptosis. Apoptosis is an orderly cell death. There are two types of apoptosis: exogenous apoptosis via death receptor pathway, and endogenous apoptosis via mitochondrial pathway. Autophagy is a self-eating process. It serves to degrade impaired organelles and malformed proteins via autophagosomes and recycle and reuse them for cell functions. Ferroptosis is a recently discovered PCD. It is a process of iron-dependent lipid peroxidation. It usually causes a dense mitochondrial membrane and the loss of glutathione peroxidase 4.

In Section 1, "Cell Death", Chapter 1 summarizes the role of epigenetic regulation of cell death in cancer. For example, the role of epigenetic regulations in apoptosis including the DNA hypermethylation on FAS receptor, BCL-2, and Apaf-1; their role in DNA hypomethylation on chromosomal stability; and their role in the histone methylation on BIM. It also discusses the role of epigenetic regulations in other forms of cell death, including necroptosis, pyroptosis, ferroptosis, NETosis, immunogenic cell death, and parthanatos. In Chapter 2, proline-rich peptide (PRP-1) is isolated from neurosecretory granules of the bovine neurohypophysis. PRP-1 has been shown to have the opposite effects on cell death in neurodegenerative diseases and cancer. It significantly reduces staurosporine-induced apoptosis of postnatal hippocampal cells, as well as doxorubicin-induced apoptosis of bone marrow monocytes and granulocytes. PRP-1 also exerts the opposite effect on the proliferation of bone marrow stromal cells obtained from normal humans and on the stromal cells isolated from the human giant-cell tumor. PRP-1 cytostatically inhibits chondrosarcoma bulk tumors but exerts a drastic cytotoxic effect on sarcomas cancer stem cells.

In Section 2, "Ferroptosis", Chapter 3 discusses the role of ferroptosis in tumorigenesis, progression, and chemoresistance of different types of leukemia including acute myeloid leukemia (AML), acute lymphoblastic leukemia (ALL), chronic myelocytic leukemia (CML), and chronic lymphocytic leukemia (CLL). Furthermore, both activators and inhibitors of ferroptosis have been identified and used to study the molecular mechanisms underlying ferroptosis, thus paving the way toward designing new therapeutic strategies for treating ferroptosis-related diseases. In Chapter 4, ferroptosis is reported to be responsible for several neurological disorders; however, the underlying mechanism is not fully elucidated. This chapter reviews the role of ferroptosis in neurological disorders. For example, the ferroptosis-related gene, the ferroptosis regulators, and their role in ferroptosis-related neurological diseases. Chapter 5 discusses the role of iron and ferroptosis in chronic diseases, including (1) its role in tumors, such as in hepatocellular carcinoma, pancreatic cancer, renal cell carcinoma, breast cancer, bladder tumor, and tumor-associated ferroptosis regulatory protein (SLC7A11, p53, NRF2, ACSL4, GPX4, FSP1); (2) its role in neurodegenerative diseases, such as Alzheimer's disease, Parkinson's disease, and amyotrophic lateral sclerosis; and (3) its role in cardiovascular and cerebrovascular disease.

In Section 3, "Autophagy", Chapter 6 presents the authors' study demonstrating that overactivation of the mTOR pathway in the liver promotes de novo lipid synthesis and eventually the formation of non-inflammatory hepatocellular carcinoma (HCC). The mechanism study reveals that persistent activation of the mTOR pathway promotes the de novo synthesis of lipids, resulting in the production of a large amount of lipid in the liver; meanwhile, it also inhibits autophagy, resulting in the inability of lipids to be removed in time and its accumulation in the liver. Accumulated lipid peroxidation is responsible for the development of HCC. In addition, the persistently activated mTOR pathway inhibits the release of exosomes. The reduced release of exosomes may impair intercellular communication, especially with immune cells, thereby making HCC more prone to invasion and metastasis with less inflammation.

Despite significant achievements in the study of cell death and its role in diverse diseases in recent years, the mechanisms underlying cell death are still not fully elucidated. It is believed that based on a further understanding of the role and mechanism of cell death in diseases, more patients will benefit from novel treatment strategies targeting cell death. This book is a useful resource in this regard.

Ke Xu
Tianjin Lung Cancer Institute,
Tianjin Medical University General Hospital,
Tianjin, China

Section 1

Cell Death

Chapter 1

Epi-Regulation of Cell Death in Cancer

Antonio Beato, Laura Della Torre, Vincenza Capone, Daniela Carannante, Gregorio Favale, Giulia Verrilli, Lucia Altucci and Vincenzo Carafa

Abstract

How do organisms regulate the correct balance between the production of "new" cells and the elimination of the "old" ones, remains an important biology issue under investigation. Cell(s) death represents a fundamental process involved in organism development and cell homeostasis, whose alteration is considered one hallmark of cancer and lead to drug resistance and consequently treatment failure. The recent re-classification of cell death has identified new molecular programs in which several proteins have a pivotal role. Several studies have highlighted a direct link between epigenetic modifications and cell death mechanisms. Different epi-modifications have been described, capable of regulating diverse key players implicated in cell death, leading to uncontrolled proliferation of cancer cells. Scientific efforts are focused on the understanding the epigenetic regulation of cell death mechanisms by developing tools and/or new epi-molecules able to overcome cell death resistance. The development of new epi-molecular tools can overcome cell death deregulation thus potentially improving the sensitivity to the anti-tumor therapies. This chapter focuses on the main epigenetic deregulations in cell death mechanisms in cancer.

Keywords: epigenetics, cell death, cancer, apoptosis, necroptosis pyroptosis, Immunogenic cell death, NETosis, parthanatos

1. Introduction

Epigenetics is the study of functionally heritable changes in the genome that occur without structural changes in the DNA sequence [1], characterizing cellular phenomena and molecular mechanisms responsible of the remodeling of a phenotype starting from a fixed structure that is determined by the genotype [2]. Epigenetic mechanisms can regulate gene expression through covalent chemical modifications, histone posttranslational modifications (PTMs), several RNA species or also through chromosomal superstructure modifications in which DNA is packaged without making any change in the DNA (in its) basic structure [3, 4]. During the past years, different types of epigenetic mechanisms have been identified (i) DNA methylation, (ii) histone modifications, and (iii) non-coding RNA (ncRNA), able to modulate gene and protein expression [3]. Epigenetic changes are the results of the action of three different enzymatic classes, (i) *writers*, able to add chemical groups on DNA, histones and proteins; (ii) *readers,* which read and identify several "signals" through their structural domains, and (iii) *erasers* involved in the removal of chemical groups.

DNA methylation is one of the most known epigenetic modifications able to repress gene transcription and expression, especially when located near the transcription start sites of genes [5]. Well-known is the crucial role of both hypermethylation of tumor suppressor and global hypomethylation of oncogenes in tumor initiation and progression [6]. PTMs, changing the histone structure, are also able to alter gene expression. [7]. These alterations, mediated by the addition of chemical groups at the N-terminal tail of histones are covalent, reversible, and redundant creating a real "histone code" that regulates the chromatin structure, gene expression and the recruitment of different enzymes. These changes can include (i) acetylation of lysine residues; (ii) methylation of lysine and arginine residues; (iii) phosphorylation of serine residues; (iv) the binding of one or more monomers of ubiquitin (ubiquitination), (v) the binding of several polypeptide (SUMOylation), (vi) citrullination, consisting in the conversion of arginine residues into citrulline residues by specific enzymes [8]. Much progress has been made in understanding the roles of both microRNAs (miRNAs) and long non-coding RNAs (lncRNAs) in gene regulation which play an important part in several biological processes such as proliferation, differentiation, apoptosis [9].

Epigenetic changes are often associated to the genesis of many pathologies with a "high social impact" such as cancer. These modifications induce the blocking or definitive silencing of many cellular signal transduction pathways, the restoration of which today represents a promising therapeutic perspective. Particularly

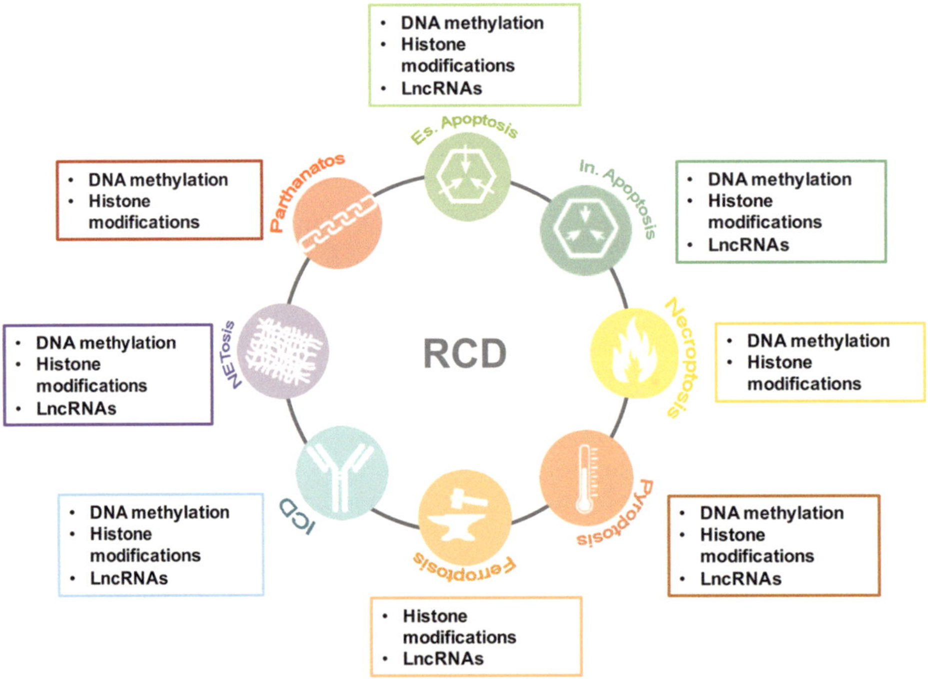

Figure 1.
Epigenetic regulation in cell deaths.

studied are the cell death pathways, complex and finely regulated processes whose deregulation alters the correct cell homeostasis, responsible of the excessive cell proliferation. Cell death *phenomena* alteration represents one of the main markers of oncogenic cell transformation responsible for resistance to cancer and drug therapies failure [10]; indeed, a tumor cell still retains the ability to proliferate and/or to go into apoptosis but the pathways of regulation of these signals can be silenced and therefore inactive.

The aim of this chapter is to shed light on the epigenetic regulation of the molecular players involved in cell death pathways, whose alteration has a pivotal role in carcinogenesis. Considering the reversibility of the epigenetic modifications, they represent a promising target for anticancer therapy (**Figure 1**).

2. Cell death mechanisms

For a long time, cell death was considered an inexorable event for cells, an indispensable cellular mechanism which allows the cell to die when it is damaged, altered or simply aged [11]. Cell death is to be considered not only as a destructive process for the organism but also as a defensive process that the cells put in place to preserve homeostasis [12]. Several studies have shown that cell death mechanisms not only allow the cell to die when it has reached the end of its life cycle, but they are useful events both during prenatal and the development for the removal of excess, damaged or altered formed cells [13]. Based on the severity of the insults, on the morphology and on the biological events that can be activated during the cell death process, we distinguish an accidental death cellular process (ACD), and a regulated cell death (RCD) [14].

Necrosis, a type of ACD, is an unordered and unscheduled death mechanism that cells puts in place in response to stimuli such as radiations, toxins, osmotic variations, viral or bacterial infections followed by a large immunogenic and inflammatory response. Some enzyme systems involved in this process are lytic enzymes called calpain and cathepsins as well as damage-associated molecular patterns (DAMPs) which can be DNA fragments, ATP, uric acid, inflammatory cytokines including High Mobility Group Box 1 (HMGB1), an inflammatory cytokine of great importance in the necrosis process [15].

On the other hand, RCD, is a controlled death process that can be genetically or pharmacologically regulated which is involved in two different scenarios. It acts as the main process in tissue development responsible for the cell's turnover in the absence of exogenous environmental perturbations [16, 17]. RCD can be also the result of prolonged intra and extracellular perturbations [18] and does not alter tissue homeostasis or cell development. When it occurs in physiological conditions it is called programmed cell death (PCD) [19–21]. Considering only morphological characteristics, it has been proposed a classification of several forms of cell death including: type I or apoptosis, type II or autophagy, type III or necrosis [14, 19].

In this last decade, a new subdivision of the various cell deaths has been proposed through essential and mechanistic aspects that distinguish them. Twelve types of cell deaths have been identified which are Necroptosis, Ferroptosis, Pyroptosis, Parthanatos, Entotic cell deah, NETotic cell death, Lysosome-dependent cell death (LDCD), Autophagy-dependent cell death (ADCD), Immunogenic cell death (ICD), Intrinsic apoptosis, Extrinsic apoptosis, Mithocondrial permeability transition-driven necrosis (MPT).

2.1 Epigenetic regulation in apoptosis

Apoptosis, or type I PCD, is a finely regulated "molecular assisted suicide" mechanism, necessary for maintaining cellular homeostasis processes. It is the response to DNA damage (spontaneous apoptosis), or different conditions such as hypoxia, lack of growth factors and action of chemotherapeutic agents (induced apoptosis). It is also involved in physiological processes such as embryogenesis and differentiation. It is defined as a "clean" death mechanism since there is no release of waste elements: the apoptotic bodies - which contain cell fragments - are eliminated through the action of the immune system and more specifically through the action of macrophages [22]. The loss of apoptotic regulation causes uncontrolled cell proliferation leading to several human diseases such as cancer [23]. Apoptosis is the result of extrinsic or intrinsic signals, coming from outside and inside the cell, respectively. A pivotal role, in this pathway, is played by initiator and effector caspases synthesized as inactive zymogens and activated by a proteolytic cut [24]. The extrinsic apoptosis pathway is triggered by the link between death receptors of the Tumor Necrosis Factor (TNF)-family with their specific pro-apoptotic ligands resulting in the activation of different molecular adapter able to cleave initiator caspases which in turn cleave and activate effector caspases [25] while the intrinsic pathway is triggered by mitochondrial dysfunction caused by cellular stress [26]. The main event is the release of cytochrome c from complex, called apoptosome, with other cytosolic proteins Apoptotic protease activating factor-1 (Apaf-1) and activates initiator and effector caspases [27].

Several epigenetic modifications have been identified as responsible for the evasion of the apoptotic process and carcinogenesis. As result of an alteration of DNA methyltransferases (DNMTs) functions, in cancer cells diffused events of hyper- and hypo-methylation, contributes to apoptosis resistance [28].

In several cancers, hypermethylation of the promoter region of tumor suppressor genes involved in the regulation of apoptotic processes leads to an uncontrollable proliferation contributing to apoptosis resistance of cancer cells [29].

Hypermethylation on FAS promoter region, is responsible of the suppression of its expression, leading to a Cutaneous T-cell lymphoma and neoplastic transformation of epithelial cells into colon cancer [30, 31]. In neuroblastoma, melanoma and ovarian cancer cells, the resistance to TRAIL-induced apoptosis is due to hypermethylation of the DR4 and DR5 promoters [32–34]. In other cancer types, such as hepatocellular carcinoma, bladder cancer, small-cell lung carcinoma, glioblastoma, retinoblastoma, and neuroblastoma, caspases 8 and 10 are silenced by the methylation on their promoters resulting in a block of apoptotic pathway [35–39]. Silencing of Apaf-1, as result of the block of intrinsic apoptotic pathway, is observed in leukemia and melanoma, as well as bladder and kidney cancers and is associated with therapeutic resistance [40–43].

Promoter hypermethylation of BAX, BAK, and PUMA in multiple myeloma and Burkitt's lymphoma cells, is responsible for the silencing of these genes and so of the abrogation of related death pathway while in prostate cancer patients, despite the hypermethylation of the Bcl-2 promoter, apoptotic pathways, particularly the extrinsic pathway, are largely preserved [44–46].

However, also a global genomic hypomethylation has a role for carcinogenesis [47]. In a variety of human cancers, including metastatic tumor, B-cell chronic lymphocytic leukemia, cervical, colorectal, hepatocellular and bladder cancer hypomethylation determine chromosomal instability and cancer transformation [48–52]. In addition to DNA methylation, other epigenetic modifications, such as histone modification and miRNA regulation, can alter apoptotic pathway. In Burkitt's lymphoma, a well-known repressive chromatin mark, the trimethylation of lysine 27 of histone H3 (H3K27me3), affects the expression levels of proapoptotic BIM protein [53]. In medulloblastoma patients, abnormal H3 and H4 acetylation patterns at the promoter region of DR4 gene expression, alter apoptosis [54]. Similarly, increased H3 and H4 acetylation induced by HDAC inhibitors affect the amounts of proapoptotic Bax protein in human colon cancer cells leading to cell cycle arrest and apoptosis [55].

An alteration of the balance between Histone Acetyltransferases (HATs) and Histone Deacetylases (HDACs) contributes to cancer promotion modulating the acetylation levels of several non-histone proteins involved in apoptotic cell death pathway such as Rb, E2F and ku70. Indeed, the involvement of Ku70 in promoting apoptosis, is strictly regulated by its acetylation level. Ku70, inhibits BAX activation, preventing its translocation to the mitochondrial membrane and suppressing apoptosis. Ku70 acetylation promoted by CBP and PCAF on two different lysine residues (K539 and K542), blocks Ku70-BAX connection and promotes apoptosis [56]. Acetylation of E2F1 is essential for the recruitment of several proteins that control the apoptotic response to DNA damage. In response to DNA damage, acetylated E2F1 interacts with Rb influencing the cellular response driven transcription of the proapoptotic target gene p73 [57, 58].

Through the regulation of gene expression, miRNAs are considered key regulators of several cellular processes such as apoptosis and have a pivotal role in cancer progression. A function in tumorigenesis was described for miR15/16 as well as for some miR-34 family members. In pituitary adenoma, B-cell chronic lymphocytic leukemia and prostate cancer, miR15/16 miRNAs down regulated or deleted led to overexpression of antiapoptotic BCL-2, as well as cyclin D1, MCL1, and WNT3A at the post-transcriptional level inducing cancer cell proliferation and invasiveness [59–62]. Further investigation indicated a positive feedback loop between p53 and miRNAs. P53 regulates miRNA expression at numerous levels and, as a transcription factor, p53 can affect the expression of individual miRNAs [63, 64].

MiR-34a and miR-34b/c, three members of the mir34 family, are direct p53 targets. MiR-34 family regulated SIRT1 mRNA leading to an increase in p53 acetylation levels which regulate cell-cycle and apoptosis [65]. MiR-34a is repressed via hypermethylation in different types of cancer such as gastric cancer, chronic lymphocytic leukemia, pancreatic, breast, colon, kidney cancer, and Burkitt's lymphoma [66], while miR-34b/c was down regulated in sarcoma, colon, and ovarian cancer [62]. Another miRNA, the miR-29b, able to target DNMT3b and MCL1 is significantly reduced in several cancers such as lung, pancreatic and ovarian [67–70]. An hypermethylation of miR-127 is characterized in cancers of the bladder, prostate, breast, and lung, as well as lymphoma [71]. This epi-modification is responsible of the miR-127 silencing, which in turn determines the hyperactivity of one of its molecular targets, the proto-oncogene BCL-6, in these cancers [72]. Other examples are miR-106b and miR-93, which are known to alter TGF-induced apoptosis in gastric cancer cells by inhibiting BIM expression while MiR-135a inhibits JAK2, resulting in a decrease in antiapoptotic Bcl-xL expression [73, 74]. MiR-135a expression is reduced in ovarian cancer, Hodgkin lymphoma, Acute Myeloid Leukemia (AML) (**Table 1**) [75, 76].

2.2 Epigenetic regulation in necroptosis

Necroptosis, a form of regulated cell death independent from caspase activation, is regulated by specific death receptors, including (but not limited to) FAS/APO-1

Cell death	Epigenetic modification	Targets	Cancers	References
Apoptosis	DNA hypermethylation	FAS receptor	CTCL, CRC	[30, 31]
		DR4, DR5	NB, Melanoma, OC	[32–34]
		Caspase 8 and Caspase 10	HCC, TCC, SCLC, GBM, Rb, NB	[35–39]
		Apaf-1	AML, Melanoma, TCC, RCC	[40–43]
		BCL-2	PDAC	[44]
		BAX, BAK, PUMA	MM, BL	[45, 46]
		miRNA 34a	GC, CLL, PDAC, BC, CRC, RCC, BL	[66]
		miRNA 34b/c	SARC, CRC, OC	[62]
		miRNA 127	TCC, PC, BC, LC, Lymphoma	[71]
	DNA hypomethylation	Chromosomal stability	CLL, CC, CRC, HCC, TCC	[48–52]
	Histone methylation	BIM	BL	[53]
	Histone acetylation and deacetylation	DR4	MB	[54]
		BAX	CRC	[55]
		Rb, E2F and Ku70	Cancer progression	[56–58]
	miRNA 15/16	BCL2, Cyclin D1, MCL1, WNT3a	Pituitary adenomas, CLL, PC	[59–62]
	miRNA 29b	DNMT3b, MCL1	NSCLC, PDAC, OC	[67–70]
	miRNA 106b and miRNA 93	BIM	GC	[73, 74]
	miRNA 135a	JAK2	OC, HL, AML	[75, 76]

Table 1.
Epigenetic regulation in apoptosis.

(CD95) and TNFR1, or pathogen recognition receptors (PRRs), including TLR3, TLR4, and Z-DNA binding protein 1 (ZBP1; also known as DAI) [77]. Necroptotic signaling pathway depends on the sequential activation of the receptor-interacting serine/threonine-protein kinase 3 (RIPK3), mixed lineage kinase domain like pseudokinase (MLKL) and (at least in some settings) on the kinase activity of RIPK1, also called necrosome [19, 78]. Therefore, it is not surprising that necroptotic cell death signaling can also be regulated by epigenetic modifications at the necrosome components [79].

Necroptosis may represent a new therapeutic strategy to overcome resistance to apoptosis. In cancer, necroptosis has been defined as a *double-edged sword* for its pro- or anti-tumor effect [80]. Epigenetic alterations may modify the gene expression levels of the necroptosis regulators, affecting cancer initiation, promotion and progression [81]. Hypo- and hyper-methylation of key components of necroptosis existed in multiple tumors and could affect gene expression and prognosis of cancer patients [81]. A multi-omics approach identified promoter hypermethylation of (i)

MLKL in skin cutaneous melanoma (SKCM) and in colon adenocarcinoma (COAD); (ii) RIPK3 in adrenocortical carcinoma (ACC); (iii) RIPK1 in kidney renal clear cell carcinoma (KIRC) and kidney renal papillary cell carcinoma (KIRP). Differently, MLKL hypomethylation has been reported in low grade glioma (LGG) and uveal melanoma (UVM); RIPK3 hypomethylation in LGG, AML and KIRC; RIPK1 in LGG, thymoma (THYM), lung squamous cell carcinoma (LUSC), ACC, and SKCM [81]. Among the necrosome components, RIPK3 is often downregulated, in cancer which is why several studies focused on its epigenetic modifications, unlike RIPK1 or MLKL.

RIPK3 is normally expressed in normal tissues, but the genomic region near the *RIPK3* transcription start site (TSS) is highly methylated resulting in loss of RIPK3 expression in different types of primary cancers probably due to an adaptive process to evade necroptosis [82, 83]. In breast cancer, 85% of patients have reduced RIPK3 expression due to promoter hypermethylation [82]. However, robust re-expression of RIPK3 in recurrent breast tumor cells was unexpectedly noted. These data were confirmed by ChIP-Seq experiments in which RNA polymerase II occupies the promoter region of *RIPK3* and epigenetic histone markers, H3K9ac and H3K4me3, are enriched in the regulatory regions of the *RIPK3* gene adjacent to the RNA polymerase II binding site. Conversely, many of the cytosines in the *RIPK3* CpG island are methylated in primary tumors. However, treatment with HDAC inhibitors and/or hypomethylating agents, such as 5-azacytidine (5-AC), can restore RIPK3 expression and thus promotes sensitivity to chemotherapeutic agents in a RIPK3-dependent manner [82, 83].

As in breast cancer, RIPK3 expression is reduced also in lung cancer and this is associated with a poorer chemotherapy response. The promoter region of *RIPK3* being highly rich in CpG island is hypermethylated differently from primary human bronchial epithelial cells. The epigenetic silencing is responsible for RIPK3 and necroptotic cell death suppressions with worse response in non-small lung cancer (NSCLC) patients receiving chemotherapy. Therefore, demethylation treatments could improve the anticancer efficacy of chemotherapy [84]. A further study investigating the epigenetic landscape of necroptosis in lung adenocarcinoma (LUAD) did not identify any correlation between the levels of methylation in the *RIPK3* promoter and its mRNA expression [85].

The role of RIPK3 has also been discussed in malignant mesothelioma (MM) as downregulation at the transcriptional level consistent with epigenetic silencing via DNA methylation was observed in 62% of primary MMs. The high frequency of CpG methylation in the *RIPK3* promoter (22%) is mediated by DNA methyltransferase DNMT1 which contributes to a very poor overall survival (OS). In human pleural MM cells, *RIPK3* gene expression decrease both *in vitro* and in primary tumors, strengthening its pivotal role as tumor suppressor [86].

Some authors identified that the methylation carried out by DNMT1 in binding to the *RIPK3* promoter is stimulated by the oncometabolite in the tricarboxylic acid (TCA) cycle, 2-hydroxyglutarate (2-HG) produced by tumor-associated isocitrate dehydrogenases 1 (IDH1) mutation [86]. Tumorigenesis could be driven by IDH1 mutation at position 132 (R132) resulting in high levels of 2-HG production, which regulates DNMT1 activity by promoting its binding to specific DNA regions including the TSS of the *RIPK3* promoter. This phenomenon investigated in human brain cancers implies resistance to necroptosis and may support the survival of cancer cells, eventually leading to tumor formation [87].

Ten Eleven Translocation (TET) methylcytosine dioxygenases enzymes, using α-ketoglutarate (α-KG) as substrate, catalyze the oxidation of 5-methylcytosine (5mC) to 5-hydroxymethylcytosine (5-hmC), which is the first step for active DNA demethylation [88]. Some intermediates of the TCA cycle, including fumarate and α-KG, can competitively inhibit the enzymatic activity of TETs [89]. In Epstein–Barr virus-encoded latent membrane protein 1 (EBV-LMP1) positive cells, high levels of fumarate and low levels of α-KG, determine *RIPK3* silencing as the result of hypermethylation of its promoter region [89]. The oncomine database refers a significant downregulation of *RIPK3* in nasopharyngeal carcinoma (NPC), compared to nasopharyngitis tissues, as the result from the impairment of TETs' enzymatic activity in EBV-LMP1 positive cells [89]. From an epigenetic regulation/modification point of view, RIPK3 (among the necrosome components) was the most investigated and promising mediator. For instance, very little is known about the epigenetic regulation of MLKL which is essential for the execution of necroptosis. Interestingly, in Burkitt's lymphoma cell lines, MLKL expression levels correlate with the methylation status. As a result of the activity of the new DNA hypomethylating agent SGI-110, the silenced expression of MLKL is restored [90].

In conclusion, these studies indicate RIPK3 as a critical regulator of necroptosis, which is considered a tumor suppressor gene and whose low expression, also regulated at the epigenetic level, can be associated with poor prognosis in cancer. Hence, treatment with hypomethylating agents alone or in combination with chemotherapeutic agents facilitate the activation of necroptotic signaling (**Table 2**).

2.3 Epigenetic regulation in pyroptosis

Pyroptosis is a form of inflammatory RCD induced by the activation of the NF-kB pathway followed by the triggering of intracellular sensors / receptors such as NLRP3, NLRC4 and AIM2, in response to DAMPs, Pathogen-associated molecular pattern (PAMPs) or different cytotoxic stimuli [91, 92]. Assembly of the inflammasome leads to pyroptotic cell death mediated by the cleavage of Gasdermin-D (GSDM-D), by caspases (caspase1 or caspase-4/5/11) and to the release of Interleukin-1*β* *(Il-1β)* and Il-18 in the microenvironment [93]. Pyroptosis can also occur with an alternative mechanism by which caspase-3 activates GSDM-E [94]. Recent studies have identified an epigenetic modulation of the pyroptotic process in cancer [95]. Among the

Cell death	Epigenetic modification	Targets	Cancers	References
Necroptosis	DNA hypermethylation	MLKL	SKCM, COAD, BL	[81, 90]
		RIPK3	ACC, BC, NSCLC, MESO, NPC	[81, 82, 84, 86, 89]
	DNA hypomethylation	RIPK1	KIRK, KIRP, LGG, THYM, LUSC, ACC, SKCM	[81]
		RIPK3	LGG, KIRK, AML	[81]
		MLKL	LGG, UVM	[81]
	Histone methylation	RIPK3	BC	[82]
	Histone acetylation	RIPK3	BC	[82]

Table 2.
Epigenetic regulation in necroptosis.

proteins involved in the pyroptosis, epigenetic modification related to NLRP3, the sensor ASC, caspase-1 and GSDMs are the best characterized [96].

In gastric carcinoma, the loss of caspase-1 gene expression, which appears to be related to the worsening of the patient's prognosis, could be associated with methylation phenomena [97]. Indeed, anticancer therapy with the 5-aza-C hypomethylating agent, activates transcriptional mechanisms with expression of caspase-1 and the conclusion of the pyroptotic program [97]. In gastric, NPC and breast cancer, the hypermethylation at the promoter of the tumor suppressor ZDHHC1, induces pyroptosis by increasing the activation of caspase-1 in response to accumulated oxidative damage [98]. DNA methylation plays a crucial role in NLRP3-inflammasome activation in human monocytes, where, under physiological conditions, Cas-1a, ASC and Il-1β promoters are hypermethylated [95]. Furthermore, NF-kB and the demethylase TET2 are responsible for the hypomethylation and reactivation of ASC and Il-1β genes in differentiated monocytes and macrophages [95]. In lung, gastric and renal cancers, the hypermethylated state of ASC increased tumor growth and is associated with poor prognosis [99], indeed some studies report ASC demethylation as a possible strategy to induce selective cell death in cancer cells [100]. Conversely, reduced methylation in the ASC promoter, often associated with migration and invasion which are the basis of the metastatic process, is reported in patients with glioblastoma and squamous cell carcinoma [101, 102]. NLRP3-acetylation is fundamental for the assembly of the ASC domain and for its activation in response to exogenous stimuli in aging-associated inflammatory diseases as cancer; thus, the NLRP3 deacetylation mediated by SIRT2 represses its activity and inflammasome formation [103]. These evidences were confirmed in Aged SIRT2-deficient mice with a high-fat diet, which showed an increase in plasma Il-18 followed by an increase in NLRP3-inflammasome activity [103]. The epigenetic regulation of pyroptosis may also depend on the action of small non-coding RNAs. It is known that several miRNAs bind to 3′-untranslated NLRP3 gene region and degrade it [104]. To confirm these evidences, it was demonstrated that during myeloid differentiation, low levels of miR-233 increases NLRP3 inflammasome transcription, accompanied by the release of pro-inflammatory cytokines in activated macrophages [105]. In addition, XLOC_000647 overexpression, an intergenic lncRNA, reduces the expression of NLRP3 in pancreatic cancer cells, playing a protective role against the starting of endothelial-mesenchymal transition (EndoMT), proliferation and metastasis formation, identifying a novel epigenetic mechanism involving the NLRP3-inflammasome in tumor progression of pancreatic cancer [106]. Additional research demonstrated the direct regulation of pro-caspase-1 by Neat1. This lncRNA can stabilize mature caspase-1 tetramers (p20: p10)2 and (p33: p10)2, promoting the assembly of the NLRP3-AIM2-inflammasomes, inducing a caspase-1-dependent pyroptosis [107]. The best characterized member of the GSDMs family, GSDM-D, appears to play a key role in NSCLC and can be regulated by methylation processes [108]. Elevated GSDM-D levels have been associated with unfavorable prognosis in lung cancer but favorable in skin cutaneous melanoma and its expression is regulated by the binding of Foxo1 on its promoter [109]. The hypermethylating activity of DNMT was also found at the GSDM-D promoter in lymphocytes natural killer, NK92 cells, in which it appears to be a critical checkpoint for the inhibition of the pyroptosis mechanism [110]. An indirect regulation occurs in colorectal cancer, where the rp1-85f18.6 knockout, a lncRNA highly expressed in CRC patients, leads to an increased pyroptosis through the cleavage of GSDM-D, suggesting a possible application of epigenetic modulators of inflammosomes for cancer therapy [111]. GSDM-E is found to be silenced in gastric, colorectal and breast cancer due to hypermethylation of CpG islands within its promoter and appears to be related to an increased risk of metastasis [112]. Epigenetic regulation of GSDM-E may also depend on small non-coding RNA activity such as miR-155-5p, which can bind 3'-UTR reducing GSDM-E expression [113]. A further regulation takes place thanks to the presence of lncRNA which have been shown to be involved in pathological processes of various diseases including cancer by regulating directly or indirectly proteins involved in the main pyroptotic pathways [113]. Recent discoveries have identified new molecules, which in turn can activate or inhibit the expression of GSDMs, regulating pyroptosis at the epigenetic level [114]. One of the most important is Decitabine (DAC), a DNMT inhibitor used in hematological cancers therapy combined with chemotherapy, which can regulate the expression of several genes. In particular DAC treatment in several tumor cell lines induces DFN5 gene up-regulation leading to an increase of GSDM-E protein expression followed by pyroptosis activation. [115]. Moreover, treatment with methyltransferase inhibitors (e.g. 5-aza-C) increases the expression of GSDM-E in cancer cell lines and also improves the efficacy of chemotherapeutic agents (e.g. doxorubicin) to trigger pyroptosis [116]. Furthermore, anti-inflammatory drugs such as dimethyl fumarate (DMF) and monomethyl fumarate (MMF) have shown the ability to increase transcription levels of DNMT3a and DNMT3b, leading to GSDM-D silencing via its promoter hypermethylation (**Table 3**) [108].

2.4 Epigenetic regulation in immunogenic cell death

Immunogenic Cell Death (ICD) is a process where dying cells activate an immunogenic response mediated by the release of DAMPs into the microenvironment, recognized by different immune cells and necessary for the immunological memory [117, 118].

DAMPs and nucleic acids released from dying cells, together with the release of chemo attractive agents in the microenvironment, contribute to increase the

Cell death	Epigenetic modification	Targets	Cancers	References
Pyroptosis	DNA hypermethylation	Caspase 1	GC	[97]
		ASC	NSCLC, GC, RCC	[99]
		GSDM-D	NSCLC,	[108]
		ZDHHC1	GC, NPC, BC	[98]
		GSDM-E	GC, CRC, BC	[112]
	DNA hypomethylation	ASC	GBM, SSC	[101, 102]
	Histone deacetylation	NLRP3	Inflammatory diseases	[103]
	miRNA 233	NLRP3	Myeloid differentiation	[105]
	XLOC_000647	NLRP3	PDAC	[106]
	Neat1	Pro-caspase-1	Unknown	[107]
	Rp1-85f18.6	Inflammosomes	CRC	[111]
	miRNA 155-5p	GSDM-E	Unknown	[113]

Table 3.
Epigenetic regulation in pyroptosis.

antigenicity of dying cells leading to the recruitment of innate immunity cells such as neutrophils and dendritic [119, 120].

Different molecular mechanisms are involved in this type of cell death, such as the UPR (Unfolded Protein Response) and autophagy as well as the release of many molecular players like Annexin 1, HMGB1, Interferons (IFNs) and different chemokines [121]. Under physiological stress, the endoplasmic reticulum (ER) activates the UPR, an evolutionarily conserved mechanism thanks to which ER chaperonins, Heat Shock Proteins (HSPs) such as HSP70 and Calreticulin (CALR) are translocated on the cell surface being an "eat me" signal for recognition by dendritic cells [122, 123].

Recently, it was demonstrated that epigenetic modifications can regulate several molecular players directly involved in ICD, supporting the idea for the development of new epigenetic drugs that can be used in cancer immunotherapy [121].

Histone and DNA methylation as well as ncRNAs are the main epigenetic modifications able to regulate targets that have a pivotal role in ICD such as HSPs, CALR, Annexin 1 and HMGB1 [121]. In lung cancer, inositol-requiring enzyme-1 (IRE1), an enzyme involved in UPR activation, is silenced by methylation. Indeed, treatment with Chaetocin, an Histone Lysine Methyltransferase (HKMT) inhibitor, determines an increment of the expression of this enzyme, suggesting that its regulation could be modulated via histone methylation (126,127). In colon and pancreatic cancer cell lines, the methylation at HSP90 promoter, related to an enhanced expression of DNA methyltransferase, inhibits its expression probably altering the immune response. The treatment with epigenetic modulators such as Zebularine, a DNMT inhibitor, can restore the immune response that leads to the induction of ICD [121, 124].

Different non-coding RNAs such as ncRNA-RB1, miR-27a and nc886, can modulate epigenetically CALR expression [121]. It has been shown that in A549 cell line (adenocarcinoma alveolar basal epithelial) the knockdown of ncRNA-RB1 reduces the expression of CALR, altering its translocation on the cell surface and probably influencing the fate of ICD [125]. Downregulation of Calreticulin was observed also in colorectal cancer by miR-27a action, resulting in a blocked Major Histocompatibility Complex (MHC) class I cell surface exposure [126]. In malignant gastric cancer cell lines, such as SNU-005, SNU-484 and MKN-01, the activity of the long non-coding RNA nc886, which has anti-proliferative and tumor suppressor roles [127, 128], has been found decreased compared to the non-malignant gastric cell line HFE-145, probably due to the CpG hypermethylation at the nc886 promoter region [128]. In nasopharyngeal carcinoma cell lines, both gene and protein expression of Annexin 1 are downregulated by methylation phenomena [129]. In head and neck squamous cell carcinoma, the presence of miRNA-196a/b epigenetically regulates Annexin 1, downregulating both mRNA and protein levels [130]. At the level of epigenetic regulation, it is thought that HMGB1 could act as an epigenetic modifier able to silence Tumor Necrosis Factor-alpha (TNF-α) and Il-1β [131]. miRNA-129-2, a tumor suppressor in glioma and hepatocellular carcinoma [132, 133], can inhibit the release of HMGB1. The regulatory region of this miRNA is strongly methylated in portions of its promoter region leading to its suppression and consequent expression of HMGB1 [134, 135]. Autophagy is essential for the ICD process as it promotes the synthesis and transport of ATP from the cell which is fundamental for an optimal immunogenic response [120, 136]. In submandibular carcinomas the expression of P2RX7 receptor is controlled by the methylation of its promoter and aberrant methylation phenomena may interfere with its expression and the related pathway [137]. Hypermethylation affects other autophagy players such as the Tensin Homolog (PTEN), as demonstrated in melanoma and in breast and stomach cancer [138, 139] and the Autophagy-Related

Protein 5 (ATG5), studied in melanoma and colorectal cancer. This epi-modification leads to a downregulation of PTEN and ATG5 protein expression during cancer progression [138–141]. The expression of CXCL10 in ovarian cancer cells, may depend on the methylation of its promoter, indeed the use of demethylating agents is able to increase its expression [142]. Acetylation can also modulate ICD [143], indeed Histone Deacetylase 3 (HDAC3)-deficient macrophages, stimulated with LPS, are unable to activate several genes involved in inflammation including IFNβ, demonstrating a main role for HDAC3 in controlling IFNβ expression. [144]. Furthermore, it has been shown that the use of caloric restriction mimetics (CRMs) may have a pivotal role in anticancer immunosurveillance [145]. CRMs stimulate ATP release by influencing acetylation of histone proteins showing a potential epigenetic mechanism able to induce or not autophagy during cancer (**Table 4**) [145, 146].

2.5 Epigenetic regulation in ferroptosis

Ferroptosis is a newly discovered form of RCD reliant on iron-dependent lipid peroxidation [149]. The increase in free iron and the accumulation of lipid peroxides occurs through the action of a small molecule called erastin which can induce non-apoptotic cell death in an ST (Small T oncoproteins) and RAS^{V12} (oncogenic allele of HRAS)-dependent way [150].

By the interaction with voltage-gated anion channels (VDAC), erastin can inhibit the cysteine/glutamate transport system X_c^- (SLC7A11) leading to cysteine depletion, glutathione deficiency with excessive lipid peroxidation and consequently induction of ferroptotic cell death [151].

Some evidences highlight an epigenetic regulation of ferroptosis. For instance, ncRNAs regulate the progression of NSCLC mediating ferroptosis [152]. P53RRA, a cytosolic lncRNA, by interacting with G3BP1, promotes ferroptosis trough the

Cell death	Epigenetic modification	Targets	Cancers	References
Immunogenic cell death	DNA hypermethylation	HSP90	CRC, PDAC	[121, 124]
		Annexin 1	NPC	[129]
		P2RX receptor	SGC	[137]
		CXCL10	OC	[142]
		PTEN	Melanoma, BC and SCr	[142, 143]
	Histone methylation	IRE-1	NSCLC	[147, 148]
	Histone acetylation	IFNβ, CRMs	Cancer progression	[144–146]
	ncRNA-RB1	CALR	LUAD	[125]
	miRNA 27a	CALR	CRC	[130]
	nc886	CALR	GC	[127, 128]
	miRNA 19a/b	Annexin 1	HNSC	[130]
	miRNA 129–2	HMGB1	Glioma, HCC	[132, 133]

Table 4.
Epigenetic regulation in immunogenic cell death.

activation of p53 pathway and the transcription of some metabolic genes responsible of the increased intracellular concentration of iron and ROS lipids and of the inhibition of growth induced by erastine [153].

In lung cancer, the nuclear lncRNA LINC00336 is upregulated and, through the interaction with ELAV-like-RNA-binding protein 1 (ELAVL1), acts as an inhibitor of ferroptosis by decreasing the intracellular levels of iron and ROS lipids. Moreover, LINC00336 also acts as an endogenous sponge for another microRNA (miR-6852) which is a negative regulator of cystathionine-β-synthase (CBS) that has a pivotal role in ferroptosis [154].

Treatment of NSCLC cell line NCI-H1299 with XAV939, a Wnt/−catenin pathway inhibitor, resulted in a downregulated SLC7A11 expression that controls iron concentration and the activation of ferroptosis-mediated pathways responsible of the suppression of NSCLC progression [152]. Furthermore, the deubiquitinase DUB, a tumor suppressor inactivated in many types of tumors [155], after the assembly of the polycomb repressive deubiquitanase complex (PR-DUB) is able to inhibit the ubiquitinated histone H2A (H2Aub) placement on the SLCA711 promoter whose down-regulation blocks ferroptosis through the cysteine starvation and GSH depletion [156]. The monoubiquitination of H2B on lysine 120 (H2Bub1), a marker of transcriptional activation involved in the regulation of the Warburg effect and tumorigenesis [157], regulates both the expression of SLC7A11 and of a group of ion-binding genes linked to metabolism classifying this modification as a new epigenetic regulator of ferroptosis [158]. The activity of Lysine Demethylase 3B (KDM3B) inhibits erastin-induced ferroptosis through the activation of SLC7A11, cooperating with the transcription factor ATF4 [159]. In addition, BRD family proteins, including BRD4, can also participate in the epigenetic regulation of ferroptosis. The use of BRD4 inhibitor JQ1 has been shown to induce ferroptosis through the downregulation of GPX4, SLC7A11 and SLC3A2 expression in breast and lung cancer cells classifying it as a potential therapeutic agent in cancer treatment (**Table 5**) [160, 161].

2.6 Epigenetic regulation in NETosis

NETosis is a form of cell death exclusive for neutrophils, caused by the uncontrolled production of netotic bodies, useful in physiological conditions for the neutralization of pathogens. The mechanism originates with the activation of ion channels associated with receptors able to modify the intracellular levels of calcium. Subsequent phosphorylation pathways lead to the production of mitochondrial ROS and the calcium-dependent activation of PAD4, responsible for the chromatin decondensation and the end of the NETotic process [162]. In breast cancer, the release of cancer extracellular chromatin networks (CECNs) into the microenvironment

Cell death	Epigenetic modification	Targets	Cancers	References
Ferroptosis	Histone demethylation	SLCA711, ATF4	Unknown	[159]
	Histone ubiquitination	SLCA711	Several cancers	[158-160]
	LINC00336	ELAVL1	LC	[154]
	XAV939	SLC7A11	NSCLC	[152]

Table 5.
Epigenetic regulation in ferroptosis.

appears to be related to the onset of lung metastases [163]. Among the key molecular processes of NETosis, the role played by the PAD4 enzyme is well known, which increases the levels of citrullination of histones in a calcium-dependent manner leading to chromatin decondensation and netotic nuclear collapse [164]. Several studies on patients with different tumors, such as breast, colorectal and lung cancer, have found an important increase in plasma levels of hypercitrullinated histone H3, suggesting it as a potential prognostic marker [165–167]. The hyper-citrullination of H3 is a widespread phenomenon during the formation of NETotic bodies as well as reduced levels of methylation of arginine 3 on histone H4 and high levels of acetylated lysine 16 on histone H4 as reported in breast cancer [163]. The increase in the enzymatic activity of PAD4 and in the netotic process is closely related to its epigenetic regulation. In MCF7 cancer cells, citrullination of the OKL38 promoter by PAD4 was described, suggesting a correlation between NETs formation and breast cancer [168]. Increased angiogenesis and deposition of fibrous material in malignant tumors also appears to be related to PAD4-mediated citrullination of antithrombin (cAT) [169]. In hepatocellular carcinomas, the global hypomethylated state of DNA and the hypermethylation of promoters of genes involved in tumorigenesis, such as p53 and p21, may partially depend on the reduced action of PAD4, on the expression and the enzymatic activity of DNMT3a [170].

In colon cancer, miR-155 can ensure the translation of PAD4 mRNA, inducing the netotic process and the tumor progression [171]. New evidence has proved the role played by miR-505 in breast and pancreatic cancer. It negatively regulates SIRT3 by altering mitochondrial metabolism and ROS production, triggering the production of NETs (**Table 6**) [172].

2.7 Epigenetic regulation in parthanatos

Parthanatos is a type of PCD characterized by hyperactivation of poly (ADP-ribose) polymerase 1 (PARP-1) followed by PAR accumulation and mitochondrial release of apoptosis inducing factor (AIF) [173]. The molecular interaction between AIF and the macrophage migration inhibitory factor (MIF) leads to massive DNA fragmentation and cell collapse [174]. The knowledge related to the epigenetic modification involved in parthanatic process and their role in tumors is currently poorly known. In liver cancer, the damage induced by

Cell death	Epigenetic modification	Targets	Cancers	References
NETosis	DNA hypermethylation	P53, p21	HCC	[170]
	DNA hypomethylation	Global DNA	HCC	[170]
	Histone methylation	H4	BC	[163]
	Histone acetylation	H4	BC	[163]
	Citrullination	H3	BC, CRC, LC	[167, 169]
		OKL38, cAT	BC	[168, 169]
	miRNA 155	PAD4	CRC	[171]
	miRNA 505	SIRT3	BC, PDAC	[172]

Table 6.
Epigenetic regulation in NETosis.

UV rays causes the activation of PARP-1 and the PARylation of histones, with the consequent recall of ALC1 on chromatin and activation of DNA repair [175]. In fact, PARP-1 can facilitate the recruitment of repair systems through the decondensation of chromatin independently by ubiquitylation [176]. In breast cancer, the behavior and function of insulators is controlled by PARP-1, through conformational changes of chromatin. The increase in PARylation of CCCTC-binding factor (CTCF), triggers its functions as an insulator, activating mechanisms able to induce DNA hypomethylation, central feature of many forms of cancer [177, 178]. The correlation between PARP-1, chromatin opening and gene transcription activation is poorly explained in the literature. In breast cancer, PARP-1 allows chromatin access to RNA-pol II with the inhibition of demethylase activity of KDM5B by PARylation, leading to the global hypomethylation of H3K4 [179]. A very recent study identifies the lysine demethylase KDM6B as a key factor in the epigenetic control of parthanatos and in the response to antitumor therapy with alkylating agents. The reduction of KDM6B levels leads to the activation of DNA repair checkpoints mediated by MGMT, causing alkylating agents resistance. Conversely, the increase in KDM6B levels favors parthanatic cell death induced by alkylating agents [180]. These new insights open the window to understanding the epigenetic mechanisms underlying parthanatos and the epigenetic function of PARP-1 (**Table 7**).

3. Conclusions

Epigenetics regulates several processes including differentiation, development, growth and cell death. Specifically, cell death controls various physiological and pathological phenomena that are crucial for life development. A deeper knowledge of both cell death and epigenetics, and their interconnections, might be the key to better understand how different processes in life are modulated and how to exploit them therapeutically.

The fact that epi-deregulation in cancer clearly also alters the main players of the different cell death pathways has important consequences. For examples, some epi compounds (i.e. HKMT, HDAC, HMT inhibitors) might be used also for regulation of the expression of the main players involved in cell death and might, in turn, help for cell death pathways reactivation in cancer or, also into the recognition of cancer cells by the immune system. In addition, the identification of possible epigenetic biomarkers linked to cell death players deregulation could be beneficial to contrast several cancers strengthening the well-known concept of "personalized therapy".

Cell death	Epigenetic modification	Targets	Cancers	References
Parthanatos	DNA hypomethylation	H3	BC	[179]
	Histone demethylation	KDM6B	Unknown	[180]
	PARylation	Histones	HCC	[175]
		CTCF	Several cancers	[177, 178]

Table 7.
Epigenetic regulation in parthanatos.

Acknowledgements

We thank D. Mancinelli for English language editing. We thank: MUR, Proof of Concept POC01_00043; POR Campania FSE 2014e2020 ASSE III. Nuovi Farmaci e Biomarkers di risposta e Resistenza Farmacologica nel cancro del ColonRetto-Nabucco no.1682, MISE; Valere 2019:CAMPANIA. Ricerca Finalizzata of Italian Ministry of Health (GR-2018-12366268).

Conflict of interest

The authors declare no conflict of interest.

Declaration

Vincenza Capone and Giulia Verrilli hold a PhD fellowship "Dottorati Innovativi con Caratterizzazione Industriale" coded: B97J20000140006 (VC); B63D21008370006 (GV).

Abbreviations

CTCL	Cutaneous T-cell lymphoma
CRC	Colorectal cancer
NB	Neuroblastoma
OC	Ovarian cancer
HCC	Hepatocellular carcinoma
TCC	Transitional cell carcinoma
SCLC	Small cell lung cancer
GBM	Glioblastoma
Rb	Retinoblastoma
AML	Acute myeloid leukemia
RCC	Renal cell carcinoma
PDAC	Pancreatic adenocarcinoma
PC	Prostate cancer
MM	Multiple myeloma
BL	Burkitt's lymphoma
GC	Gastric cancer
CLL	Cronic lymphocytic leukemia
BC	Breast cancer
CC	Cervical cancer
MB	Medulloblastoma
NSCLC	Non-small cell lung cancer
HL	Hodgkin lymphoma
SKCM	Skin cutaneous melanoma
COAD	Colon adenocarcinoma
ACC	Adrenocortical carcinoma
KIRK	Kidney renal clear cell carcinoma
KIRP	Kidney renal papillary cell carcinoma

LGG	Low grade glioma
UVM	Uveal melanoma
THYM	Thymoma
LUSC	Lung squamous cell carcinoma
STAD	Stomach adenocarcinoma
SCC	Squamous cell carcinoma
SGC	Salivary gland cancer
HNSC	Head and neck squamous cell carcinoma
LC	Lung cancer
SARC	Sarcoma
SCr	Stomach cancer
MESO	Mesothelioma

Author details

Antonio Beato[1], Laura Della Torre[1], Vincenza Capone[1], Daniela Carannante[1], Gregorio Favale[1], Giulia Verrilli[1], Lucia Altucci[1,2,3*†] and Vincenzo Carafa[1*†]

1 Università degli Studi della Campania "L. Vanvitelli", Naples, Italy

2 Biogem, Molecular Biology and Genetics Research Institute, Ariano Irpino, Italy

3 IEOS CNR, Napoli, Italy

*Address all correspondence to: lucia.altucci@unicampania.it
and vincenzo.carafa@unicampania.it

† Co-last authors.

IntechOpen

© 2022 The Author(s). Licensee IntechOpen. This chapter is distributed under the terms of the Creative Commons Attribution License (http://creativecommons.org/licenses/by/3.0), which permits unrestricted use, distribution, and reproduction in any medium, provided the original work is properly cited.

References

[1] Zhang L, Lu Q, Chang C. Epigenetics in health and disease. Advances in Experimental Medicine and Biology. 2020;**1253**:3-55

[2] Deans C, Maggert KA. What do you mean, "epigenetic"? Genetics. 2015;**199**(4):887-896

[3] Al Aboud NM, Tupper C, Jialal I. Genetics, Epigenetic Mechanism. 2022 Aug 8. In: StatPearls [Internet]. Treasure Island (FL): StatPearls Publishing; 2022

[4] Skvortsova K, Iovino N, Bogdanović O. Functions and mechanisms of epigenetic inheritance in animals. Nature Reviews. Molecular Cell Biology. 2018;**19**(12):774-790

[5] Jones PA. Functions of DNA methylation: Islands, start sites, gene bodies and beyond. Nature Reviews. Genetics. 2012;**13**(7):484-492

[6] Ehrlich M. DNA hypermethylation in disease: Mechanisms and clinical relevance. Epigenetics. 2019;**14**(12):1141-1163

[7] Zhang Y, Sun Z, Jia J, Du T, Zhang N, Tang Y, et al. Overview of histone modification. Advances in Experimental Medicine and Biology. 2021;**1283**:1-16

[8] Wang R, Xin M, Li Y, Zhang P, Zhang M. The Functions of histone modification enzymes in cancer. Current Protein & Peptide Science. 2016;**17**(5):438-445

[9] Dostie J, Mourelatos Z, Yang M, Sharma A, Dreyfuss G. Numerous microRNPs in neuronal cells containing novel microRNAs. RNA. 2003;**9**(2):180-186

[10] Labi V, Erlacher M. How cell death shapes cancer. Cell Death & Disease. 2015;**6**:e1675

[11] Orrenius S, Nicotera P, Zhivotovsky B. Cell death mechanisms and their implications in toxicology. Toxicological Sciences. 2011;**119**(1):3-19

[12] Gudipaty SA, Conner CM, Rosenblatt J, Montell DJ. Unconventional ways to live and die: Cell death and survival in development, homeostasis, and disease. Annual Review of Cell and Developmental Biology. 2018;**34**:311-332

[13] Singh R, Letai A, Sarosiek K. Regulation of apoptosis in health and disease: The balancing act of BCL-2 family proteins. Nature Reviews. Molecular Cell Biology. 2019;**20**(3): 175-193

[14] della Torre L, Nebbioso A, Stunnenberg HG, Martens JHA, Carafa V, Altucci L. The role of necroptosis: Biological relevance and its involvement in cancer. Cancers. 2021;**13**(4):684

[15] Lotfi R, Kaltenmeier C, Lotze MT, Bergmann C. Until death do us part: Necrosis and oxidation promote the tumor microenvironment. Transfusion Medicine and Hemotherapy. 2016;**43**(2):120-132

[16] Fuchs Y, Steller H. Programmed cell death in animal development and disease. Cell. 2011;**147**(4):742-758

[17] Conradt B. Genetic control of programmed cell death during animal development. Annual Review of Genetics. 2009;**43**:493-523

[18] Galluzzi L, Bravo-San Pedro JM, Kepp O, Kroemer G. Regulated cell death and adaptive stress responses. Cellular and Molecular Life Sciences. 2016;**73**(11-12):2405-2410

[19] Galluzzi L, Vitale I, Abrams JM, Alnemri ES, Baehrecke EH, Blagosklonny M v, et al. Molecular definitions of cell death subroutines: Recommendations of the nomenclature committee on cell death 2012. Cell Death and Differentiation 2012;19(1):107-120

[20] Bidan CM, Fratzl M, Coullomb A, Moreau P, Lombard AH, Wang I, et al. Magneto-active substrates for local mechanical stimulation of living cells. Scientific Reports. 2018;**8**(1):1464

[21] Tang D, Kang R, Berghe T, vanden Vandenabeele P, Kroemer G. The molecular machinery of regulated cell death. Cell Research. 2019;**29**(5):347-364

[22] Elmore S. Apoptosis: A review of programmed cell death. Toxicologic Pathology. 2007;**35**(4):495-516

[23] Jan R, Chaudhry GES. Understanding apoptosis and apoptotic pathways targeted cancer therapeutics. Advanced Pharmaceutical Bulletin. 2019;**9**(2):205-218

[24] Shi Y. Caspase activation, inhibition, and reactivation: A mechanistic view. Protein Science. 2004;**13**(8):1979-1987

[25] Sayers TJ. Targeting the extrinsic apoptosis signaling pathway for cancer therapy. Cancer Immunology, Immunotherapy. 2011;**60**(8):1173-1180

[26] Wang SF, Chen S, Tseng LM, Lee HC. Role of the mitochondrial stress response in human cancer progression. Experimental Biology and Medicine. 2020;**245**(10):861-878

[27] Bratton SB, Salvesen GS. Regulation of the Apaf-1-caspase-9 apoptosome. Journal of Cell Science. 2010;**123**(Pt 19): 3209-3214

[28] Roll JD, Rivenbark AG, Jones WD, Coleman WB. DNMT3b overexpression contributes to a hypermethylator phenotype in human breast cancer cell lines. Molecular Cancer. 2008;**7**:15

[29] Ozyerli-Goknar E, Bagci-Onder T. Epigenetic deregulation of apoptosis in cancers. Cancers. 2021;**13**(13):3210

[30] Wu J, Wood GS. Reduction of Fas/CD95 promoter methylation, upregulation of Fas protein, and enhancement of sensitivity to apoptosis in cutaneous T-cell lymphoma. Archives of Dermatology. 2011;**147**(4):443-449

[31] Petak I, Danam RP, Tillman DM, Vernes R, Howell SR, Berczi L, et al. Hypermethylation of the gene promoter and enhancer region can regulate Fas expression and sensitivity in colon carcinoma. Cell Death and Differentiation. 2003;**10**(2):211-217

[32] van Noesel MM, van Bezouw S, Voûte PA, Herman JG, Pieters R, Versteeg R. Clustering of hypermethylated genes in neuroblastoma. Genes, Chromosomes & Cancer. 2003;**38**(3):226-233

[33] Bae SI, Cheriyath V, Jacobs BS, Reu FJ, Borden EC. Reversal of methylation silencing of Apo2L/TRAIL receptor 1 (DR4) expression overcomes resistance of SK-MEL-3 and SK-MEL-28 melanoma cells to interferons (IFNs) or Apo2L/TRAIL. Oncogene. 2008;**27**(4):490-498

[34] Horak P, Pils D, Haller G, Pribill I, Roessler M, Tomek S, et al. Contribution of epigenetic silencing of tumor necrosis factor-related apoptosis inducing ligand receptor 1 (DR4) to TRAIL resistance and ovarian cancer. Molecular Cancer Research. 2005;**3**(6):335-343

[35] Cho S, Lee JH, Cho SB, Yoon KW, Park SY, Lee WS, et al. Epigenetic methylation and expression of caspase

8 and survivin in hepatocellular carcinoma. Pathology International. 2010;**60**(3):203-211

[36] Malekzadeh K, Sobti RC, Nikbakht M, Shekari M, Hosseini SA, Tamandani DK, et al. Methylation patterns of Rb1 and Casp-8 promoters and their impact on their expression in bladder cancer. Cancer Investigation. 2009;**27**(1):70-80

[37] Shivapurkar N, Toyooka S, Eby MT, Huang CX, Sathyanarayana UG, Cunningham HT, et al. Differential inactivation of caspase-8 in lung cancers. Cancer Biology & Therapy. **1**(1):65-69

[38] Hervouet E, Vallette FM, Cartron PF. Impact of the DNA methyltransferases expression on the methylation status of apoptosis-associated genes in glioblastoma multiforme. Cell Death & Disease. 2010;**1**:e8

[39] Harada K, Toyooka S, Shivapurkar N, Maitra A, Reddy JL, Matta H, et al. Deregulation of caspase 8 and 10 expression in pediatric tumors and cell lines. Cancer Research. 2002;**62**(20):5897-5901

[40] Furukawa Y, Sutheesophon K, Wada T, Nishimura M, Saito Y, Ishii H, et al. Methylation silencing of the Apaf-1 gene in acute leukemia. Molecular Cancer Research. 2005;**3**(6):325-334

[41] Soengas MS, Capodieci P, Polsky D, Mora J, Esteller M, Opitz-Araya X, et al. Inactivation of the apoptosis effector Apaf-1 in malignant melanoma. Nature. 2001;**409**(6817):207-211

[42] Christoph F, Kempkensteffen C, Weikert S, Köllermann J, Krause H, Miller K, et al. Methylation of tumor suppressor genes APAF-1 and DAPK-1 and in vitro effects of demethylating agents in bladder and kidney cancer. British Journal of Cancer. 2006;**95**(12):1701-1707

[43] Jia L, Srinivasula SM, Liu FT, Newland AC, Fernandes-Alnemri T, Alnemri ES, et al. Apaf-1 protein deficiency confers resistance to cytochrome c-dependent apoptosis in human leukemic cells. Blood. 2001;**98**(2):414-421

[44] Carvalho JR, Filipe L, Costa VL, Ribeiro FR, Martins AT, Teixeira MR, et al. Detailed analysis of expression and promoter methylation status of apoptosis-related genes in prostate cancer. Apoptosis. 2010;**15**(8):956-965

[45] Pompeia C, Hodge DR, Plass C, Wu YZ, Marquez VE, Kelley JA, et al. Microarray analysis of epigenetic silencing of gene expression in the KAS-6/1 multiple myeloma cell line. Cancer Research. 2004;**64**(10):3465-3473

[46] Garrison SP, Jeffers JR, Yang C, Nilsson JA, Hall MA, Rehg JE, et al. Selection against PUMA gene expression in Myc-driven B-cell lymphomagenesis. Molecular and Cellular Biology. 2008;**28**(17):5391-5402

[47] Feinberg AP, Vogelstein B. Hypomethylation distinguishes genes of some human cancers from their normal counterparts. Nature. 1983;**301**(5895):89-92

[48] Bedford MT, van Helden PD. Hypomethylation of DNA in pathological conditions of the human prostate. Cancer Research. 1987;**47**(20):5274-5276

[49] Wahlfors J, Hiltunen H, Heinonen K, Hämäläinen E, Alhonen L, Jänne J. Genomic hypomethylation in human chronic lymphocytic leukemia. Blood. 1992;**80**(8):2074-2080

[50] Kim YI, Giuliano A, Hatch KD, Schneider A, Nour MA, Dallal GE, et al.

Global DNA hypomethylation increases progressively in cervical dysplasia and carcinoma. Cancer. 1994;**74**(3):893-899

[51] Esteller M. Epigenetics in cancer. The New England Journal of Medicine. 2008;**358**(11):1148-1159

[52] Ehrlich M. DNA methylation in cancer: Too much, but also too little. Oncogene. 2002;**21**(35):5400-5413

[53] Paschos K, Smith P, Anderton E, Middeldorp JM, White RE, Allday MJ. Epstein-barr virus latency in B cells leads to epigenetic repression and CpG methylation of the tumor suppressor gene Bim. PLoS Pathogens. 2009;**5**(6):e1000492

[54] Aguilera DG, Das CM, Sinnappah-Kang ND, Joyce C, Taylor PH, Wen S, et al. Reactivation of death receptor 4 (DR4) expression sensitizes medulloblastoma cell lines to TRAIL. Journal of Neuro-Oncology. 2009;**93**(3):303-318

[55] Myzak MC, Dashwood WM, Orner GA, Ho E, Dashwood RH. Sulforaphane inhibits histone deacetylase in vivo and suppresses tumorigenesis in Apc-minus mice. The FASEB Journal. 2006;**20**(3):506-508

[56] Cohen HY, Lavu S, Bitterman KJ, Hekking B, Imahiyerobo TA, Miller C, et al. Acetylation of the C terminus of Ku70 by CBP and PCAF controls Bax-mediated apoptosis. Molecular Cell. 2004;**13**(5):627-638

[57] Ianari A, Gallo R, Palma M, Alesse E, Gulino A. Specific role for p300/CREB-binding protein-associated factor activity in E2F1 stabilization in response to DNA damage. The Journal of Biological Chemistry. 2004;**279**(29):30830-30835

[58] Glozak MA, Sengupta N, Zhang X, Seto E. Acetylation and deacetylation of non-histone proteins. Gene. 2005;**363**:15-23

[59] Cimmino A, Calin GA, Fabbri M, Mv I, Ferracin M, Shimizu M, et al. miR-15 and miR-16 induce apoptosis by targeting BCL2. Proceedings of the National Academy of Sciences of the United States of America. 2005;**102**(39):13944-13949

[60] Bottoni A, Piccin D, Tagliati F, Luchin A, Zatelli MC, degli Uberti EC. miR-15a and miR-16-1 down-regulation in pituitary adenomas. Journal of Cellular Physiology. 2005;**204**(1):280-285

[61] Bonci D, Coppola V, Musumeci M, Addario A, Giuffrida R, Memeo L, et al. The miR-15a-miR-16-1 cluster controls prostate cancer by targeting multiple oncogenic activities. Nature Medicine. 2008;**14**(11):1271-1277

[62] Vogt M, Munding J, Grüner M, Liffers ST, Verdoodt B, Hauk J, et al. Frequent concomitant inactivation of miR-34a and miR-34b/c by CpG methylation in colorectal, pancreatic, mammary, ovarian, urothelial, and renal cell carcinomas and soft tissue sarcomas. Virchows Archiv. 2011;**458**(3):313-322

[63] Liao JM, Cao B, Zhou X, Lu H. New insights into p53 functions through its target microRNAs. Journal of Molecular Cell Biology. 2014;**6**(3):206-213

[64] Raver-Shapira N, Oren M. Tiny actors, great roles: microRNAs in p53's service. Cell Cycle. 2007;**6**(21):2656-2661

[65] Yamakuchi M, Ferlito M, Lowenstein CJ. miR-34a repression of SIRT1 regulates apoptosis. Proceedings of the National Academy of Sciences of the United States of America. 2008;**105**(36):13421-13426

[66] Lodygin D, Tarasov V, Epanchintsev A, Berking C, Knyazeva T,

Körner H, et al. Inactivation of miR-34a by aberrant CpG methylation in multiple types of cancer. Cell Cycle. 2008;**7**(16):2591-2600

[67] Garzon R, Liu S, Fabbri M, Liu Z, Heaphy CEA, Callegari E, et al. MicroRNA-29b induces global DNA hypomethylation and tumor suppressor gene reexpression in acute myeloid leukemia by targeting directly DNMT3A and 3B and indirectly DNMT1. Blood. 2009;**113**(25):6411-6418

[68] Yanaihara N, Caplen N, Bowman E, Seike M, Kumamoto K, Yi M, et al. Unique microRNA molecular profiles in lung cancer diagnosis and prognosis. Cancer Cell. 2006;**9**(3):189-198

[69] Wang LH, Huang J, Wu CR, Huang LY, Cui J, Xing ZZ, et al. Downregulation of miR-29b targets DNMT3b to suppress cellular apoptosis and enhance proliferation in pancreatic cancer. Molecular Medicine Reports. 2018;**17**(2):2113-2120

[70] Flavin R, Smyth P, Barrett C, Russell S, Wen H, Wei J, et al. miR-29b expression is associated with disease-free survival in patients with ovarian serous carcinoma. International Journal of Gynecological Cancer. 2009;**19**(4):641-647

[71] Saito Y, Liang G, Egger G, Friedman JM, Chuang JC, Coetzee GA, et al. Specific activation of microRNA-127 with downregulation of the proto-oncogene BCL6 by chromatin-modifying drugs in human cancer cells. Cancer Cell. 2006;**9**(6):435-443

[72] Chen J, Wang M, Guo M, Xie Y, Cong YS. miR-127 regulates cell proliferation and senescence by targeting BCL6. PLoS One. 2013;**8**(11):e80266

[73] Petrocca F, Visone R, Onelli MR, Shah MH, Nicoloso MS, de Martino I, et al. E2F1-regulated microRNAs impair TGFbeta-dependent cell-cycle arrest and apoptosis in gastric cancer. Cancer Cell. 2008;**13**(3):272-286

[74] Wu H, Huang M, Cao P, Wang T, Shu Y, Liu P. MiR-135a targets JAK2 and inhibits gastric cancer cell proliferation. Cancer Biology & Therapy. 2012;**13**(5):281-288

[75] Duan S, Dong X, Hai J, Jiang J, Wang W, Yang J, et al. MicroRNA-135a-3p is downregulated and serves as a tumor suppressor in ovarian cancer by targeting CCR2. Biomedicine & Pharmacotherapy. 2018;**107**:712-720

[76] Xu H, Wen Q. Downregulation of miR-135a predicts poor prognosis in acute myeloid leukemia and regulates leukemia progression via modulating HOXA10 expression. Molecular Medicine Reports. 2018;**18**(1):1134-1140

[77] Zhai J, Hinton TM, Waddington LJ, Fong C, Tran N, Mulet X, et al. Lipid-PEG conjugates sterically stabilize and reduce the toxicity of phytantriol-based lyotropic liquid crystalline nanoparticles. Langmuir. 2015;**31**(39):10871-10880

[78] Choi ME, Price DR, Ryter SW, Choi AMK. Necroptosis: A crucial pathogenic mediator of human disease. JCI Insight. 2019;**4**(15):128834

[79] Berger SL, Kouzarides T, Shiekhattar R, Shilatifard A. An operational definition of epigenetics. Genes & Development. 2009;**23**(7):781-783

[80] Wang T, Jin Y, Yang W, Zhang L, Jin X, Liu X, et al. Necroptosis in cancer: An angel or a demon? Tumor Biology. 2017;**39**(6):1010428317711539

[81] Li G, Wang X, Liu Y, Li H, Mu H, Zhang Y, et al. Multi-omics analysis reveals the panoramic picture of necroptosis-related regulators in pan-cancer. Aging. 2022;**14**(12):5034-5058

[82] Koo GB, Morgan MJ, Lee DG, Kim WJ, Yoon JH, Koo JS, et al. Methylation-dependent loss of RIP3 expression in cancer represses programmed necrosis in response to chemotherapeutics. Cell Research. 2015;**25**(6):707-725

[83] Lin CC, Mabe NW, Lin YT, Yang WH, Tang X, Hong L, et al. RIPK3 upregulation confers robust proliferation and collateral cystine-dependence on breast cancer recurrence. Cell Death and Differentiation. 2020;**27**(7):2234-2247

[84] Wang Q, Wang P, Zhang L, Tessema M, Bai L, Xu X, et al. Epigenetic regulation of RIP3 suppresses necroptosis and increases resistance to chemotherapy in nonsmall cell lung cancer. Translational Oncology. 2020;**13**(2):372-382

[85] Zhao C, Xiong K, Adam A, Ji Z, Li X. Necroptosis identifies novel molecular phenotypes and influences tumor immune microenvironment of lung adenocarcinoma. Frontiers in Immunology. 2022;**13**:934494

[86] Tan Y, Sementino E, Cheung M, Peri S, Menges CW, Kukuyan AM, et al. Somatic epigenetic silencing of RIPK3 inactivates necroptosis and contributes to chemoresistance in malignant mesothelioma. Clinical Cancer Research. 2021;**27**(4):1200-1213

[87] Yang Z, Jiang B, Wang Y, Ni H, Zhang J, Xia J, et al. 2-HG inhibits necroptosis by stimulating DNMT1-dependent hypermethylation of the rip3 promoter. Cell Reports. 2017;**19**(9):1846-1857

[88] Koivunen P, Laukka T. The TET enzymes. Cellular and Molecular Life Sciences. 2018;**75**(8):1339-1348

[89] Shi F, Zhou M, Shang L, Du Q, Li Y, Xie L, et al. EBV(LMP1)-induced metabolic reprogramming inhibits necroptosis through the hypermethylation of the RIP3 promoter. Theranostics. 2019;**9**(9):2424-2438

[90] Koch A, Jeiler B, Roedig J, van Wijk SJL, Dolgikh N, Fulda S. Smac mimetics and TRAIL cooperate to induce MLKL-dependent necroptosis in Burkitt's lymphoma cell lines. Neoplasia. 2021;**23**(5):539-550

[91] Sagulenko V, Thygesen SJ, Sester DP, Idris A, Cridland JA, Vajjhala PR, et al. AIM2 and NLRP3 inflammasomes activate both apoptotic and pyroptotic death pathways via ASC. Cell Death and Differentiation. 2013;**20**(9):1149-1160

[92] Zhang P, Liu Y, Hu L, Huang K, Hong M, Wang Y, et al. NLRC4 inflammasome-dependent cell death occurs by a complementary series of three death pathways and determines lethality in mice. Science Advances. 2021;**7**(43):eabi9471

[93] Kv S, Deng M, JPY T. The NLRP3 inflammasome: Molecular activation and regulation to therapeutics. Nature Reviews. Immunology. 2019;**19**(8):477-489

[94] Jiang M, Qi L, Li L, Li Y. The caspase-3/GSDME signal pathway as a switch between apoptosis and pyroptosis in cancer. Cell Death Discovery. 2020;**6**:112

[95] Raneros AB, Bernet CR, Flórez AB, Suarez-Alvarez B. An epigenetic insight into NLRP3 inflammasome activation in inflammation-related processes. Biomedicine. 2021;**9**(11):1614

[96] Poli G, Fabi C, Bellet MM, Costantini C, Nunziangeli L, Romani L, et al. Epigenetic mechanisms of inflammasome regulation. International Journal of Molecular Sciences. 2020;**21**(16)

[97] do Jee C, Lee HS, Bae SI, Yang HK, Lee YM, Rho MS, et al. Loss of caspase-1 gene expression in human gastric carcinomas and cell lines. International Journal of Oncology. 2005;**26**(5):1265-1271

[98] Le X, Mu J, Peng W, Tang J, Xiang Q, Tian S, et al. DNA methylation downregulated ZDHHC1 suppresses tumor growth by altering cellular metabolism and inducing oxidative/ ER stress-mediated apoptosis and pyroptosis. Theranostics. 2020;**10**(21):9495-9511

[99] Martinez R, Schackert G, Esteller M. Hypermethylation of the proapoptotic gene TMS1/ASC: Prognostic importance in glioblastoma multiforme. Journal of Neuro-Oncology. 2007;**82**(2):133-139

[100] Salminen A, Kauppinen A, Hiltunen M, Kaarniranta K. Epigenetic regulation of ASC/TMS1 expression: Potential role in apoptosis and inflammasome function. Cellular and Molecular Life Sciences. 2014;**71**(10):1855-1864

[101] Wu L, Zhang C, Wang X, Ding X, Deng J, Liang H. Methylation of ASC/ TMS1 promoter is associated with poor prognosis of patients with gastric cancer. Clinical & Translational Oncology. 2016;**18**(3):296-303

[102] Stone AR, Bobo W, Brat DJ, Devi NS, van Meir EG, Vertino PM. Aberrant methylation and down-regulation of TMS1/ASC in human glioblastoma. The American Journal of Pathology. 2004;**165**(4):1151-1161

[103] He M, Chiang HH, Luo H, Zheng Z, Qiao Q, Wang L, et al. An acetylation Switch of the NLRP3 inflammasome regulates aging-associated chronic inflammation and insulin resistance. Cell Metabolism. 2020;**31**(3):580-591.e5

[104] Gao L, Jiang Z, Han Y, Li Y, Yang X. Regulation of pyroptosis by ncRNA: A novel research direction. Frontiers in Cell and Development Biology. 2022;**10**:840576

[105] Bauernfeind F, Rieger A, Schildberg FA, Knolle PA, Schmid-Burgk JL, Hornung V. NLRP3 inflammasome activity is negatively controlled by miR-223. Journal of Immunology. 2012;**189**(8):4175-4181

[106] Hu H, Wang Y, Ding X, He Y, Lu Z, Wu P, et al. Long non-coding RNA XLOC_000647 suppresses progression of pancreatic cancer and decreases epithelial-mesenchymal transition-induced cell invasion by down-regulating NLRP3. Molecular Cancer. 2018;**17**(1):18

[107] Zhang P, Cao L, Zhou R, Yang X, Wu M. The lncRNA Neat1 promotes activation of inflammasomes in macrophages. Nature Communications. 2019;**10**(1):1495

[108] Muhammad JS, Jayakumar MN, Elemam NM, Venkatachalam T, Raju TK, Hamoudi RA, et al. Gasdermin D hypermethylation inhibits pyroptosis and LPS-Induced IL-1β release from NK92 cells. ImmunoTargets and Therapy. 2019;**8**:29-41

[109] Xu S, Wang J, Zhong J, Shao M, Jiang J, Song J, et al. CD73 alleviates GSDMD-mediated microglia pyroptosis in spinal cord injury through PI3K/ AKT/Foxo1 signaling. Clinical and Translational Medicine. 2021;**11**(1):e269

[110] Li Z, Ji S, Jiang ML, Xu Y, Zhang CJ. The regulation and modification of

GSDMD signaling in diseases. Frontiers in Immunology. 2022;**13**:893912

[111] Ma Y, Chen Y, Lin C, Hu G. Biological functions and clinical significance of the newly identified long non-coding RNA RP1-85F18.6 in colorectal cancer. Oncology Reports. 2018;**40**(5):2648-2658

[112] de Schutter E, Croes L, Ibrahim J, Pauwels P, op de Beeck K, Vandenabeele P, et al. GSDME and its role in cancer: From behind the scenes to the front of the stage. International Journal of Cancer. 2021;**148**(12):2872-2883

[113] Xu W, Song C, Wang X, Li Y, Bai X, Liang X, et al. Downregulation of miR-155-5p enhances the anti-tumor effect of cetuximab on triple-negative breast cancer cells via inducing cell apoptosis and pyroptosis. Aging. 2021;**13**(1):228-240

[114] Bourdonnay E, Henry T. Transcriptional and Epigenetic Regulation of Gasdermins. Journal of Molecular Biology. 2022;**434**(4):167253

[115] Fan JX, Deng RH, Wang H, Liu XH, Wang XN, Qin R, et al. Epigenetics-based tumor cells pyroptosis for enhancing the immunological effect of chemotherapeutic nanocarriers. Nano Letters. 2019;**19**(11):8049-8058

[116] Xia Y, Jin Y, Cui D, Wu X, Song C, Jin W, et al. Antitumor effect of simvastatin in combination with DNA methyltransferase inhibitor on gastric cancer via gsdme-mediated pyroptosis. Frontiers in Pharmacology. 2022;**13**:860546

[117] Zhao X, Yang K, Zhao R, Ji T, Wang X, Yang X, et al. Inducing enhanced immunogenic cell death with nanocarrier-based drug delivery systems for pancreatic cancer therapy. Biomaterials. 2016;**102**:187-197

[118] Dv K, Garg AD, Kaczmarek A, Krysko O, Agostinis P, Vandenabeele P. Immunogenic cell death and DAMPs in cancer therapy. Nature Reviews. Cancer. 2012;**12**(12):860-875

[119] Garg AD, Vandenberk L, Fang S, Fasche T, van Eygen S, Maes J, et al. Pathogen response-like recruitment and activation of neutrophils by sterile immunogenic dying cells drives neutrophil-mediated residual cell killing. Cell Death and Differentiation. 2017;**24**(5):832-843

[120] Galluzzi L, Buqué A, Kepp O, Zitvogel L, Kroemer G. Immunogenic cell death in cancer and infectious disease. Nature Reviews. Immunology. 2017;**17**(2):97-111

[121] Cruickshank B, Giacomantonio M, Marcato P, McFarland S, Pol J, Gujar S. Dying to be noticed: Epigenetic regulation of immunogenic cell death for cancer immunotherapy. Frontiers in Immunology. 2018;**9**:654

[122] Rufo N, Garg AD, Agostinis P. The unfolded protein response in immunogenic cell death and cancer immunotherapy. Trends Cancer. 2017;**3**(9):643-658

[123] Ma Y, Hendershot LM. ER chaperone functions during normal and stress conditions. Journal of Chemical Neuroanatomy. 2004;**28**(1-2):51-65

[124] Nagaraju GP, Wu C, Merchant N, Chen Z, Lesinski GB, El-Rayes BF. Epigenetic effects of inhibition of heat shock protein 90 (HSP90) in human pancreatic and colon cancer. Cancer Letters. 2017;**402**:110-116

[125] Musahl AS, Huang X, Rusakiewicz S, Ntini E, Marsico A,

Kroemer G, et al. A long non-coding RNA links calreticulin-mediated immunogenic cell removal to RB1 transcription. Oncogene. 2015;**34**(39):5046-5054

[126] Colangelo T, Polcaro G, Ziccardi P, Pucci B, Muccillo L, Galgani M, et al. Proteomic screening identifies calreticulin as a miR-27a direct target repressing MHC class I cell surface exposure in colorectal cancer. Cell Death & Disease. 2016;**7**:e2120

[127] Kunkeaw N, Jeon SH, Lee K, Johnson BH, Tanasanvimon S, Javle M, et al. Cell death/proliferation roles for nc886, a non-coding RNA, in the protein kinase R pathway in cholangiocarcinoma. Oncogene. 2013;**32**(32):3722-3731

[128] Lee KS, Park JL, Lee K, Richardson LE, Johnson BH, Lee HS, et al. nc886, a non-coding RNA of anti-proliferative role, is suppressed by CpG DNA methylation in human gastric cancer. Oncotarget. 2014;**5**(11):3944-3955

[129] Tan SX, Hu RC, Dai AG, Tang CE, Yi H, Cheng AL, et al. DNA methylation inhibites ANXA1 gene expression in nasopharyngeal carcinoma cell lines*. Progress in Biochemistry and Biophysics. 2009;**36**(10):1319-1326

[130] Álvarez-Teijeiro S, Menéndez ST, Villaronga MÁ, Pena-Alonso E, Rodrigo JP, Morgan RO, et al. Annexin A1 down-regulation in head and neck squamous cell carcinoma is mediated via transcriptional control with direct involvement of miR-196a/b. Scientific Reports. 2017;**7**(1):6790

[131] Guo ZS, Liu Z, Bartlett DL, Tang D, Lotze MT. Life after death: Targeting high mobility group box 1 in emergent cancer therapies. American Journal of Cancer Research. 2013;**3**(1):1-20

[132] Yang Y, Huang JQ, Zhang X, Shen LF. MiR-129-2 functions as a tumor suppressor in glioma cells by targeting HMGB1 and is down-regulated by DNA methylation. Molecular and Cellular Biochemistry. 2015;**404**(1-2):229-239

[133] Zhai J, Qu S, Li X, Zhong J, Chen X, Qu Z, et al. miR-129 suppresses tumor cell growth and invasion by targeting PAK5 in hepatocellular carcinoma. Biochemical and Biophysical Research Communications. 2015;**464**(1):161-167

[134] He DX, Gu F, Gao F, Hao J-J, Gong D, Gu XT, et al. Genome-wide profiles of methylation, microRNAs, and gene expression in chemoresistant breast cancer. Scientific Reports. 2016;**6**(1):24706

[135] Lujambio A, Calin GA, Villanueva A, Ropero S, Sánchez-Céspedes M, Blanco D, et al. A microRNA DNA methylation signature for human cancer metastasis. Proceedings of the National Academy of Sciences of the United States of America. 2008;**105**(36):13556-13561

[136] Galluzzi L, Pietrocola F, Bravo-San Pedro JM, Amaravadi RK, Baehrecke EH, Cecconi F, et al. Autophagy in malignant transformation and cancer progression. The EMBO Journal. 2015;**34**(7):856-880

[137] Shin YH, Kim M, Kim N, Choi SK, Namkoong E, Choi SY, et al. Epigenetic alteration of the purinergic type 7 receptor in salivary epithelial cells. Biochemical and Biophysical Research Communications. 2015;**466**(4):704-710

[138] García JM, Silva J, Peña C, Garcia V, Rodríguez R, Cruz MA, et al. Promoter methylation of the PTEN gene is a common molecular change in breast cancer. Genes, Chromosomes & Cancer. 2004;**41**(2):117-124

[139] Yin L, Cai WJ, Liu CX, Chen YZ, Hu JM, Jiang JF, et al. Analysis of PTEN methylation patterns in soft tissue sarcomas by MassARRAY spectrometry. PLoS One. 2013;**8**(5):e62971

[140] Cho DH, Jo YK, Kim SC, Park IJ, Kim JC. Down-regulated expression of ATG5 in colorectal cancer. Anticancer Research. 2012;**32**(9):4091-4096

[141] Liu H, He Z, von Rütte T, Yousefi S, Hunger RE, Simon HU. Down-regulation of autophagy-related protein 5 (ATG5) contributes to the pathogenesis of early-stage cutaneous melanoma. Science Translational Medicine. 2013;**5**(202):202ra123

[142] Peng D, Kryczek I, Nagarsheth N, Zhao L, Wei S, Wang W, et al. Epigenetic silencing of TH1-type chemokines shapes tumor immunity and immunotherapy. Nature. 2015;**527**(7577):249-253

[143] Venkateswaran K, Verma A, Bhatt AN, Shrivastava A, Manda K, Raj HG, et al. Emerging Roles of Calreticulin in Cancer: Implications for Therapy. Current Protein & Peptide Science. 2018;**19**(4):344-357

[144] Chen X, Barozzi I, Termanini A, Prosperini E, Recchiuti A, Dalli J, et al. Requirement for the histone deacetylase Hdac3 for the inflammatory gene expression program in macrophages. Proceedings of the National Academy of Sciences of the United States of America. 2012;**109**(42):E2865-E2874

[145] Pietrocola F, Pol J, Vacchelli E, Rao S, Enot DP, Baracco EE, et al. Caloric Restriction Mimetics Enhance Anticancer Immunosurveillance. Cancer Cell. 2016;**30**(1):147-160

[146] Mariño G, Pietrocola F, Eisenberg T, Kong Y, Malik SA, Andryushkova A, et al. Regulation of autophagy by cytosolic acetyl-coenzyme A. Molecular Cell. 2014;**53**(5):710-725

[147] Liu X, Guo S, Liu X, Su L. Chaetocin induces endoplasmic reticulum stress response and leads to death receptor 5-dependent apoptosis in human non-small cell lung cancer cells. Apoptosis. 2015;**20**(11):1499-1507

[148] Cherblanc FL, Chapman KL, Brown R, Fuchter MJ. Chaetocin is a nonspecific inhibitor of histone lysine methyltransferases. Nature Chemical Biology. 2013;**9**(3):136-137

[149] Chen X, Li J, Kang R, Klionsky DJ, Tang D. Ferroptosis: Machinery and regulation. Autophagy. 2021;**17**(9):2054-2081

[150] Dolma S, Lessnick SL, Hahn WC, Stockwell BR. Identification of genotype-selective antitumor agents using synthetic lethal chemical screening in engineered human tumor cells. Cancer Cell. 2003;**3**(3):285-296

[151] Dixon SJ, Lemberg KM, Lamprecht MR, Skouta R, Zaitsev EM, Gleason CE, et al. Ferroptosis: An iron-dependent form of nonapoptotic cell death. Cell. 2012;**149**(5):1060-1072

[152] Yu H, Han Z, Xu Z, An C, Xu L, Xin H. RNA sequencing uncovers the key long non-coding RNAs and potential molecular mechanism contributing to XAV939-mediated inhibition of non-small cell lung cancer. Oncology Letters. 2019;**17**(6):4994-5004

[153] Mao C, Wang X, Liu Y, Wang M, Yan B, Jiang Y, et al. A G3BP1-Interacting lncRNA promotes ferroptosis and apoptosis in cancer via nuclear sequestration of p53. Cancer Research. 2018;**78**(13):3484-3496

[154] Wang M, Mao C, Ouyang L, Liu Y, Lai W, Liu N, et al. Long noncoding RNA

LINC00336 inhibits ferroptosis in lung cancer by functioning as a competing endogenous RNA. Cell Death and Differentiation. 2019;**26**(11):2329-2343

[155] Carbone M, Yang H, Pass HI, Krausz T, Testa JR, Gaudino G. BAP1 and cancer. Nature Reviews. Cancer. 2013;**13**(3):153-159

[156] Fan L, Yin S, Zhang E, Hu H. Role of p62 in the regulation of cell death induction. Apoptosis. 2018;**23**(3-4):187-193

[157] Jing YY, Cai FF, Zhang L, Han J, Yang L, Tang F, et al. Epigenetic regulation of the Warburg effect by H2B monoubiquitination. Cell Death and Differentiation. 2020;**27**(5):1660-1676

[158] Wang Y, Yang L, Zhang X, Cui W, Liu Y, Sun QR, et al. Epigenetic regulation of ferroptosis by H2B monoubiquitination and p53. EMBO Reports. 2019;**20**(7):e47563

[159] Wang Y, Zhao Y, Wang H, Zhang C, Wang M, Yang Y, et al. Histone demethylase KDM3B protects against ferroptosis by upregulating SLC7A11. FEBS Open Bio. 2020;**10**(4):637-643

[160] Sui S, Zhang J, Xu S, Wang Q, Wang P, Pang D. Ferritinophagy is required for the induction of ferroptosis by the bromodomain protein BRD4 inhibitor (+)-JQ1 in cancer cells. Cell Death & Disease. 2019;**10**(5):331

[161] Wu Y, Zhang S, Gong X, Tam S, Xiao D, Liu S, et al. The epigenetic regulators and metabolic changes in ferroptosis-associated cancer progression. Molecular Cancer. 2020;**19**(1):39

[162] Thiam HR, Wong SL, Wagner DD, Waterman CM. Cellular mechanisms of NETosis. Annual Review of Cell and Developmental Biology. 2020;**36**:191-218

[163] Shi L, Yao H, Liu Z, Xu M, Tsung A, Wang Y. Endogenous PAD4 in breast cancer cells mediates cancer extracellular chromatin network formation and promotes lung metastasis. Molecular Cancer Research. 2020;**18**(5):735-747

[164] Leshner M, Wang S, Lewis C, Zheng H, Chen XA, Santy L, et al. PAD4 mediated histone hypercitrullination induces heterochromatin decondensation and chromatin unfolding to form neutrophil extracellular trap-like structures. Frontiers in Immunology. 2012;**3**:307

[165] Thålin C, Lundström S, Seignez C, Daleskog M, Lundström A, Henriksson P, et al. Citrullinated histone H3 as a novel prognostic blood marker in patients with advanced cancer. PLoS One. 2018;**13**(1):e0191231

[166] Ronchetti L, Terrenato I, Ferretti M, Corrado G, Goeman F, Donzelli S, et al. Circulating cell free DNA and citrullinated histone H3 as useful biomarkers of NETosis in endometrial cancer. Journal of Experimental & Clinical Cancer Research. 2022;**41**(1):151

[167] Arpinati L, Shaul ME, Kaisar-Iluz N, Mali S, Mahroum S, Fridlender ZG. NETosis in cancer: A critical analysis of the impact of cancer on neutrophil extracellular trap (NET) release in lung cancer patients vs. mice. Cancer Immunology, Immunotherapy. 2020;**69**(2):199-213

[168] Yao H, Li P, Venters BJ, Zheng S, Thompson PR, Pugh BF, et al. Histone Arg modifications and p53 regulate the expression of OKL38, a mediator of apoptosis. The Journal of Biological Chemistry. 2008;**283**(29):20060-20068

[169] Chang X, Han J, Pang L, Zhao Y, Yang Y, Shen Z. Increased PADI4 expression in blood and tissues of patients with malignant tumors. BMC Cancer. 2009;**9**:40

[170] Zhang C, Fan L, Fan T, Wu D, Gao L, Ling Y, et al. Decreased PADI4 mRNA association with global hypomethylation in hepatocellular carcinoma during HBV exposure. Cell Biochemistry and Biophysics. 2013;**65**(2):187-195

[171] Hawez A, Al-Haidari A, Madhi R, Rahman M, Thorlacius H. MiR-155 regulates PAD4-dependent formation of neutrophil extracellular traps. Frontiers in Immunology. 2019;**10**:2462

[172] Bause AS, Haigis MC. SIRT3 regulation of mitochondrial oxidative stress. Experimental Gerontology. 2013;**48**(7):634-639

[173] Zhou Y, Liu L, Tao S, Yao Y, Wang Y, Wei Q, et al. Parthanatos and its associated components: Promising therapeutic targets for cancer. Pharmacological Research. 2021;**163**:105299

[174] Fatokun AA, Dawson VL, Dawson TM. Parthanatos: Mitochondrial-linked mechanisms and therapeutic opportunities. British Journal of Pharmacology. 2014;**171**(8):2000-2016

[175] Pines A, Vrouwe MG, Marteijn JA, Typas D, Luijsterburg MS, Cansoy M, et al. PARP1 promotes nucleotide excision repair through DDB2 stabilization and recruitment of ALC1. The Journal of Cell Biology. 2012;**199**(2):235-249

[176] Luijsterburg MS, Lindh M, Acs K, Vrouwe MG, Pines A, van Attikum H, et al. DDB2 promotes chromatin decondensation at UV-induced DNA damage. The Journal of Cell Biology. 2012;**197**(2):267-281

[177] Farrar D, Rai S, Chernukhin I, Jagodic M, Ito Y, Yammine S, et al. Mutational analysis of the poly(ADP-ribosyl)ation sites of the transcription factor CTCF provides an insight into the mechanism of its regulation by poly(ADP-ribosyl)ation. Molecular and Cellular Biology. 2010;**30**(5):1199-1216

[178] Docquier F, Farrar D, D'Arcy V, Chernukhin I, Robinson AF, Loukinov D, et al. Heightened expression of CTCF in breast cancer cells is associated with resistance to apoptosis. Cancer Research. 2005;**65**(12):5112-5122

[179] Krishnakumar R, Kraus WL. PARP-1 regulates chromatin structure and transcription through a KDM5B-dependent pathway. Molecular Cell. 2010;**39**(5):736-749

[180] Yang M, Wang C, Zhou M, Bao L, Wang Y, Kumar A, et al. KDM6B promotes PARthanatos via suppression of O6-methylguanine DNA methyltransferase repair and sustained checkpoint response. Nucleic Acids Research. 2022;**50**(11):6313-6331

Chapter 2

Diverse Effects of Hypothalamic Proline-Rich Peptide (PRP-1) on Cell Death in Neurodegenerative and Cancer Diseases

Silva Abrahamyan and Karina Galoian

Abstract

The proline-rich peptide (PRP-1) isolated from neurosecretory granules of the bovine neurohypophysis, produced by N.supraopticus and N.paraventricularis, has many potentially beneficial biological effects. PRP-1 has been shown to have the opposite effects on cell death in neurodegenerative and cancer diseases. It significantly reduces staurosporine-induced apoptosis of postnatal hippocampal cells, as well as doxorubicin-induced apoptosis of bone marrow monocytes and granulocytes, in both time- and dose-dependent manner. PRP-1 also exerts the opposite effect on the proliferation of bone marrow stromal cells obtained from normal humans and on the stromal cells isolated from human giant-cell tumor. PRP-1 cytostatically inhibits chondrosarcoma bulk tumor but exerts drastic cytotoxic effect on sarcomas cancer stem cells. The same peptide caused cell death through apoptosis in rats with Ehrlich Ascites Carcinoma model.

Keywords: proline-rich peptide (PRP-1), neurodegeneration, cancer diseases, ehrlich ascites carcinoma (EAC), chondrosarcoma

1. Introduction

Our long-term scientific work has been aimed at studying the protective effects of certain physiologically active compounds, including proline-rich peptide (PRP-1, comprised of 15 amino acids residues, AGAPEPAEPAQPGVY, with an apparent molecular mass of 1475.25 Da) on brain plasticity in rats with different neurodegenerative models. While displaying neuroprotective role in those models, in cancer related studies, on the other hand, PRP-1 displayed its beneficial anti-proliferative effect in cellular context dependent manner by triggering cell death leading to drastic decrease of cancer stem cell population responsible for disease relapse and drug resistance.

2. Effect of PRP-1 on cell death in the neurodegenerative diseases

The protection of neurons from damage and death in neurodegenerative disorders, such as Alzheimer disease (AD), ischemic insults, Parkinson disease (PD), is a major challenge for neuroscientists in the twenty first century. Much attention is focused on the discovery of novel biomarkers for diagnosis and therapy of neurodegenerative diseases.

The new family of peptide neurohormones consisting of 10–15 amino acid residues isolated by prof. Galoyan and coworkers from bovine and human neurohypophysis neurosecretory granules are synthesized in the form of a common precursor protein (neurophysin-vasopressin associated glycoprotein) [1, 2]. These five peptides contain a high proportion of proline residues and, therefore, were designated as proline rich peptides (PRPs). The most studied is the bovine PRP-1 (also known as galarmin). The polypeptide is not species-specific; hence it is active in mice, humans, and rats. It has been shown that PRP-1, as well as its synthetic analogue, has many beneficial biological effects.

Neuronal injuries have been suggested to promote PRP-1 synthesis and its release as a messenger, modulating the signaling cascades and, therefore, contributing to protection, regeneration, and repair of the neurons.

The neuroprotective [3–7] and immunoregulatory [8–10] effects of PRP-1 through its involvement in the neuro-immuno-hematopoietic interaction have been demonstrated

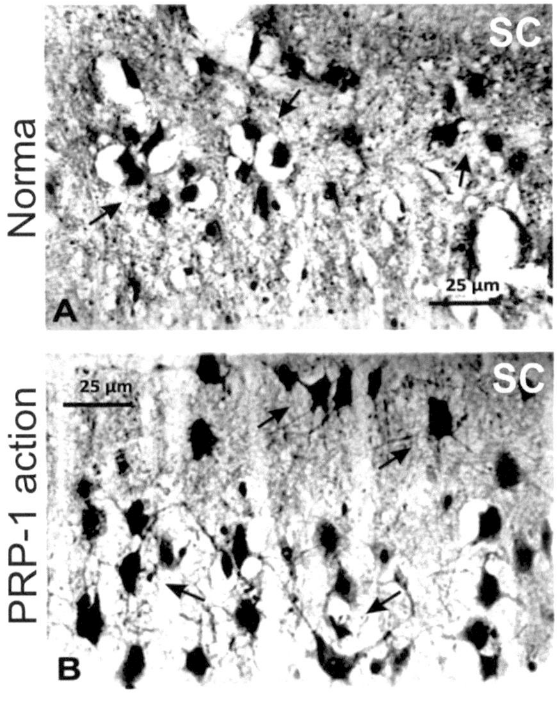

Figure 1.
PRP-1-Ir Spinal Cord (SC) nerve structures in SC-hemisectioned rats with and without PRP-1 injection. Degenerative and strongly immunoreactive for PRP-1 motoneurons (MNs) with no processes are demonstrated in the spinal cord anterior horn, situated in the lightened pericellular area, perhaps, brain edema. After daily administration of PRP-1 to trauma-injured animals for 3 weeks, regeneration of strongly immunostained MNs and their processes was observed suggested the possibility of PRP-1 involvement in the mechanisms of neuronal repair: growth of nerve fibers and motoneurons survival. ABC immunohistochemical method.

in the following models of central nervous system damage: spinal cord hemisection; βA-peptide injection (model of Alzheimer disease); vestibular nuclei damage through vibration and unilateral labyrinthectomy; immobilization stress (IMO stress).

Our previous report on the effects of PRP-1 on SC-injured rats indicated the possibility of PRP-1 involvement in the mechanisms of neuronal repair [3]. Immunohistochemical study demonstrated that treatment with PRP-1 resulted in the recovery and growth of nerve fibers, glia proliferation, and motoneuron survival. Therefore, PRP-1 has been found to be a highly active neurotrophic-like substance (**Figure 1**).

PRP-1 participation in the regeneration of the nerve structures was also immunohistochemically demonstrated in the trauma-injured rats with SC hemisection treated

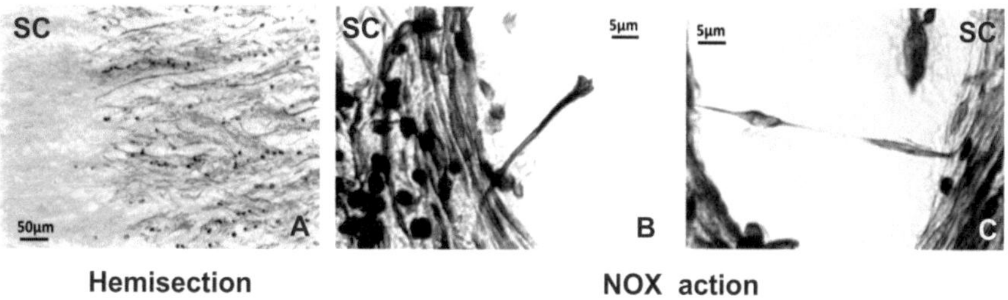

Figure 2.
SC nerve fibers in the SC hemisectioned rats regularly treated with NOX venom. Naja Naja Oxiana (NOX) snake venom prevented the scar formation, well observed two months after SC injury in the control rats (A) and resulted in the regeneration of nerve fibers growing through the trauma region (B, C). Histochemical method on detection of Ca2+-dependent acidic phosphatase activity.

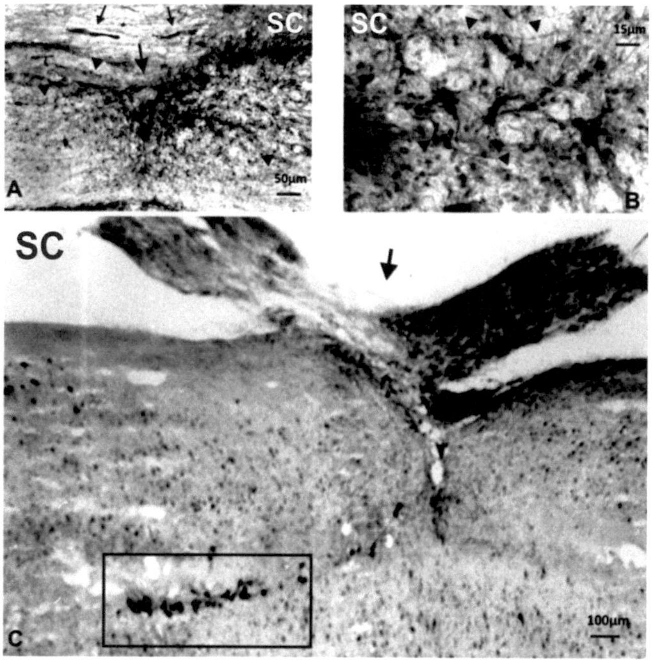

Figure 3.
PRP-1-Ir structures in SC of injured rats treated with NOX venom. NOX increased the number of PRP-1-Ir nerve fibers (A) and astrocytes (B) in the SC lesion region and promoted the survival of the PRP-1-Ir motoneurons (C). In the boxed area the PRP-1-Ir astrocytes are seen migrating toward the injured side. ABC immunohistochemical method.

by the administration of Naja Naja Oxiana (NOX) snake venom. NOX venom prevented the scar formation, well observed 2 months after the SC injury in the control rats, and resulted in the regeneration of the nerve fibers growing through the trauma region. It was suggested to exert the neuroprotective effect by involving the endogenous PRP-1 in the underlying mechanism of the neuronal recovery – based on the data regarding the survival of the immunoreactive to PRP-1 (PRP-1-Ir) motoneurons, the increased number of PRP-1-Ir nerve fibers in the SC lesion region, and appearance in the white matter of the PRP-1-Ir astrocytes migrating towards the injured side [7] (**Figures 2** and **3**).

Different forms of injury and types of stress induce different morphological responses. For example, 5-hour IMO stress induces deeper neurodegenerative changes, which is manifested by the histochemical method on detection of Ca^{2+}-dependent acid phosphatase activity. The final product – phosphate precipitate of various sizes and shapes was localized in both the neurons and the extracellular region (**Figure 4**).

In the next neurodegenerative model, the vestibular nuclei injury caused by vibration and unilateral labyrinthectomy, in various brain regions of rats, including hypothalamus, hippocampus, brain stem (locus coureleus, nucleus hyppoglosus),

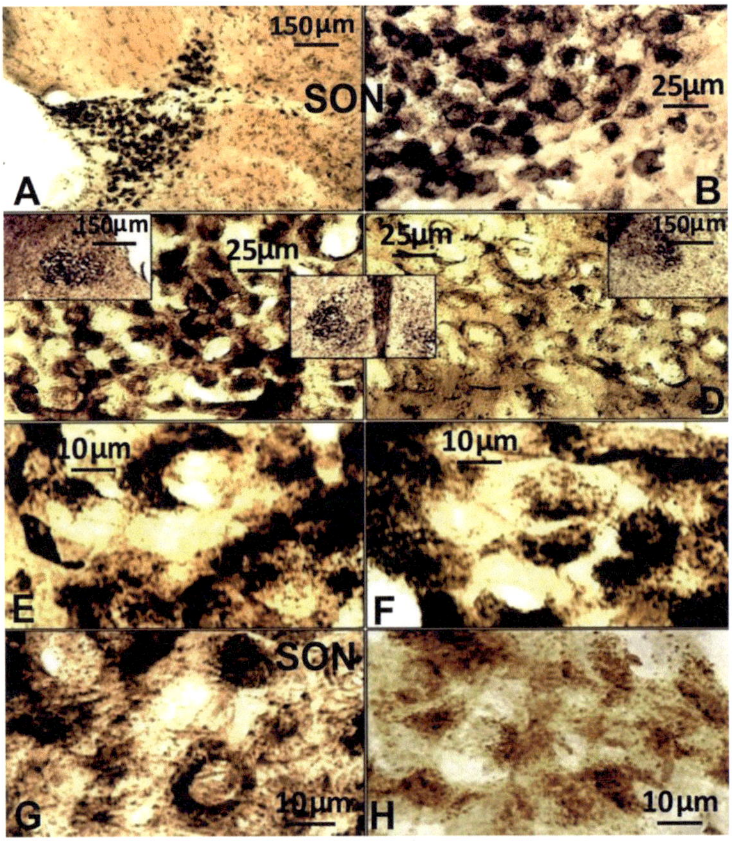

Figure 4.
Degenerative nerve structures in the hypothalamic SON and PVN of rats exposed to 5 h IMO stress. In the hypothalamic SON and PVN, the central chromatolysis and ectopied negative nuclei are revealed in the hypertrophied cells with no processes. High phosphatase activity is seen like a thich ring under the cellular membrane and extracellular area. Histochemical method on detection of Ca2+-dependent acidic phosphatase activity.

cerebellum, the neurodegeneration was revealed by the presence of the hypertrophied cells situated in the tissue edema [11]. Single PRP-1 administration triggered regeneration and survival of neurons in the same brain regions. Interestingly, obvious regeneration of the structures is revealed in labyrinthectomized rats exposed to the 2 h daily vibration for 2 weeks.

Apart from this, in the distinct stress-related brain regions of labyrinthectomized rats, neurons, demonstrating PRP-1-immunoreactivity, (PRP-1-IR) were revealed in the cell nuclei, as opposed to the PRP-1-IR in the cytoplasm of the intact cells (**Figure 5**).

According to the literature, the nuclear localization of the proteins such as c-Fos, c-Jun, as well as the so-called Heat Shock factors with a molecular weight of 1–70 kD was detected in rats under stressful conditions (exposure to radiation, temperature, chemicals, etc.) [12, 13].

Using the immediate early gene c-fos-antibody, the stress-induced activation of neurons was immunohistochemically demonstrated by detection of the c-fos-immunoreactive (c-fos-Ir) nuclei in the distinct stress-related brain regions of labyrinthectomized rats. We assume that PRP-1 can control the DNA transcription and can function as a transcription factor similar to c-fos [14].

In the same stress-related brain regions of rats, high phosphatase (APh) activity was also detected in cellular nuclei in 15 minutes after stress, earlier than gene c-fos, which can be explained by the activation of cellular activators like c-fos through their phosphorylation (**Figure 6**).

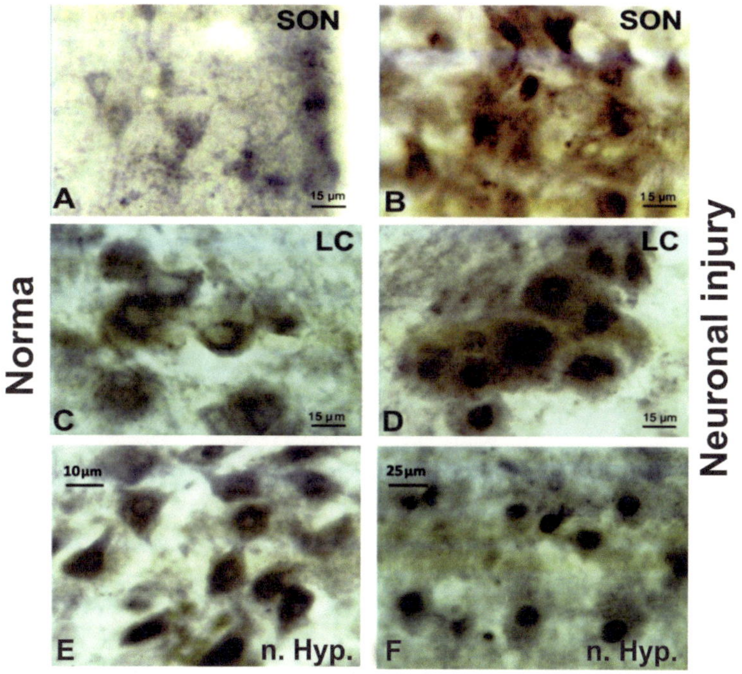

Figure 5.
PRP-1-Ir neural structures in the (A, B) supraoptic nucleus (SON), (C, D) locus coureleus (LC) and (E, F) nucleus Hyppoglosus (n.Hyp.) of intact and labyrintectomized rats brain. In the distinct stress-related brain regions of labyrintectomized rats, the stress-induced activation of neurons was found out by detection of the increased number of cell nuclei demonstrating PRP-1-IR and immediated early gene c-fos-IR. ABC immunohistochemical method.

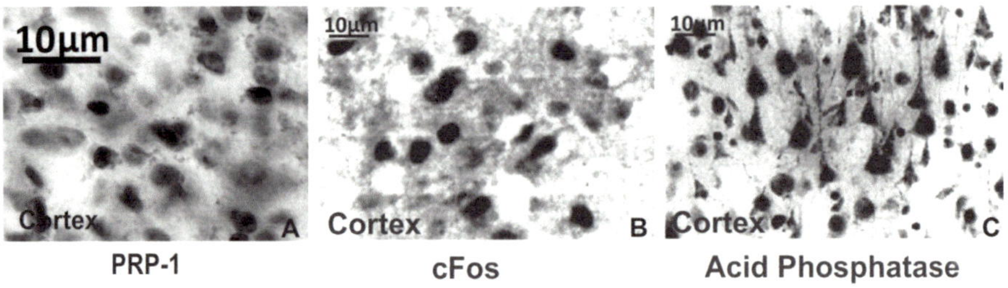

Figure 6.
PRP-1-IR, transcription factor early gene c-fos-IR (c-fos-IR) and Acid Phospatase (APh) activity in the cortex of labyrintectomized rats. Nuclei of pyramidal cells in the brain cortex of labyrintectomized rats, demonstrated (A) PRP-1- and (B) c-fos-IR, as well as (C) high APh activity. Detection of APh activity in the injured cells nuclei can be explained by activating of cellular activators like c-fos through their phosphorylation. ABC immunohistochemical method and histochemical method on detection of Ca2+-dependent acidic phosphatase activity.

Our results obtained in the mentioned models of the central nervous system injury indicate that PRP-1 therapy may protect against the neurodegeneration by enhancing the survival of the damaged neurons.

3. PRP-1 participation in the generating of new neurons

Adult brain cell regeneration, also known as neurogenesis, demonstrated in many species, including rodents, is the process of generating new neurons, [15].

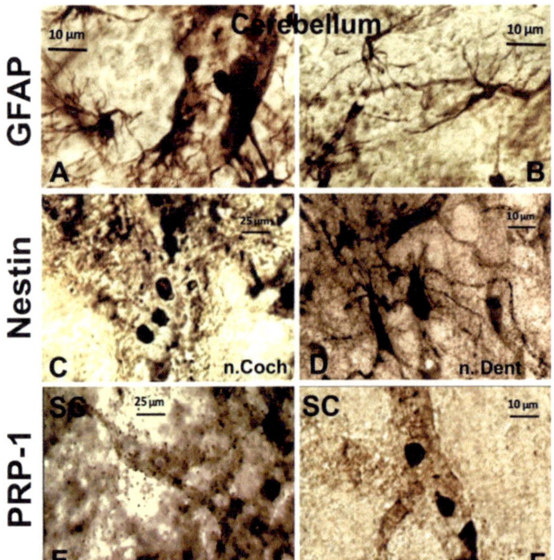

Figure 7.
GFAP-, Nestin- and PRP-1-Immunoreactive structures in the brain of 45-days aged rats exposed to prenatal IMO stress. Using markers against the neural progenitor cells, GFAP- and nestin-Ir radial astrocytes (A, B, D), nestin- and PRP-1-Ir cell structures of different sizes and forms (C, F), as well as PRP-1-Ir varicose nerve fibers and varicosities (E) were detected in the distinct stress-related brain regions. Taking into account the results obtained, we suggest that they could be the intermediate neural progenitor cells and that PRP-1 could participate in the generation of new functional neurons following the injury. ABC immunohistochemical method.

Today, there is scientific evidence of the stem cells presence in many more tissues and organs. One of their characteristics is ability to self-renew and to differentiate, to secure primary steady state functioning of a cell, called homeostasis, and, with limitations, to replace cells that die because of injury or disease [16, 17]. The list of adult tissues reported to contain stem cells is growing and includes bone marrow, peripheral blood, brain, spinal cord, muscle, and other tissues. Multipotent adult progenitor cells (MAPCs), derived from pluripotent mesenchymal stem cells, purified and isolated by Jiang et al. [18], could differentiate both into mesenchymal and neural cells.

Our recent histochemical and immunohistochemical studies in newborn rats exposed to the acute prenatal immobilization (psychogenic) stress brought new information about PRP-1 participation in the brain recovery process through generating new neurons [19].

To differentiate PRP-1-immunoreactive (PRP-1-Ir) glial cells and small undifferentiated cells found in the injured brain, we used antibodies against the astrocyte marker GFAP, neuroepithelial stem cells marker gene nestin [20, 21], and mouse stem cells. Considering the obtained results regarding the detection of GFAP-, nestin-, and

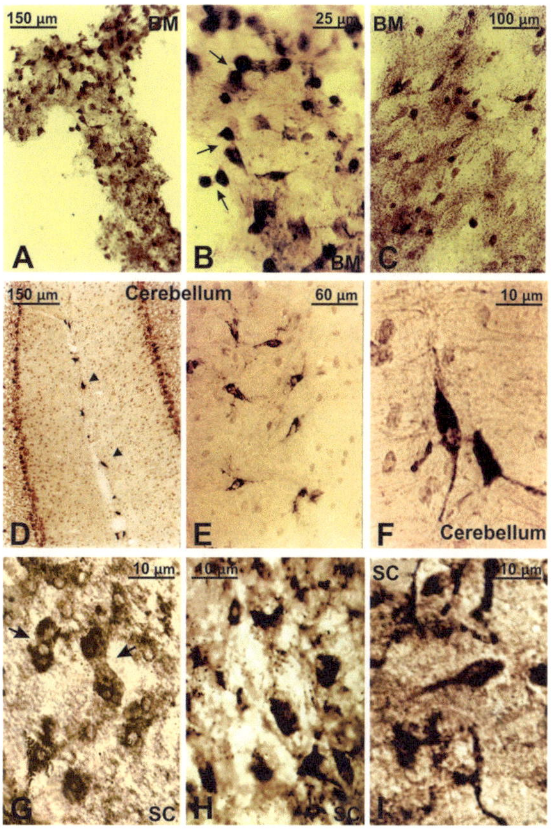

Figure 8.
Cellular structures resembling mesenchymal cells in the bone marrow (BM) and brain of the immobilized rats. (A-C) numerous round in shape and fusiform small cells with short axon-like extensions, being in the various proliferative stages (arrows), are visible in the BM stroma (B). (D) small cells (arrow heads) and (E, F) dark-colored fusiform cells with processes are detected in the cerebellum (G-I). Using antibodies against the mouse stem cells (MSCs), nestin and PRP-1, in the spinal cord of the injured rats, (G) MSCs-Ir, (H) PRP-1-Ir and (I) nestin-Ir cellular structures are revealed. Histochemical method on detection of Ca+2-dependant acid phosphatase activity (A-E) and ABC Immunohistochemical method (G-I).

PRP-1-immunoreactive radial astrocytes and cell structures of different sizes and forms, we suggest that they could be the intermediate neural progenitor cells, and that PRP-1 could participate in the generation of new functional neurons following the injury (**Figure 7**).

We succeeded also in detecting the cellular structures resembling the mesenchymal cells in both bone marrow and different stress-related brain regions of the immobilized rats some of which expressed immunoreactivity against MSCs, nestin and PRP-1 (**Figure 8**).

In addition to the established functions of PRP-1 for cell survival and neurogenesis, using the PRP-1-antiserum, as well as antiserum against the synaptophysin (presynaptic vesicle protein, marker for the functional synapsis), we also suggested that PRP-1 could mediate higher brain activity through the formation of new synapses, thus, increasing the number of connections between the neurons (**Figure 9**).

We suggest that the process of generating new neurons occurs in injured brain, most probably proceeding from BM-derived cells that migrate into the brain and

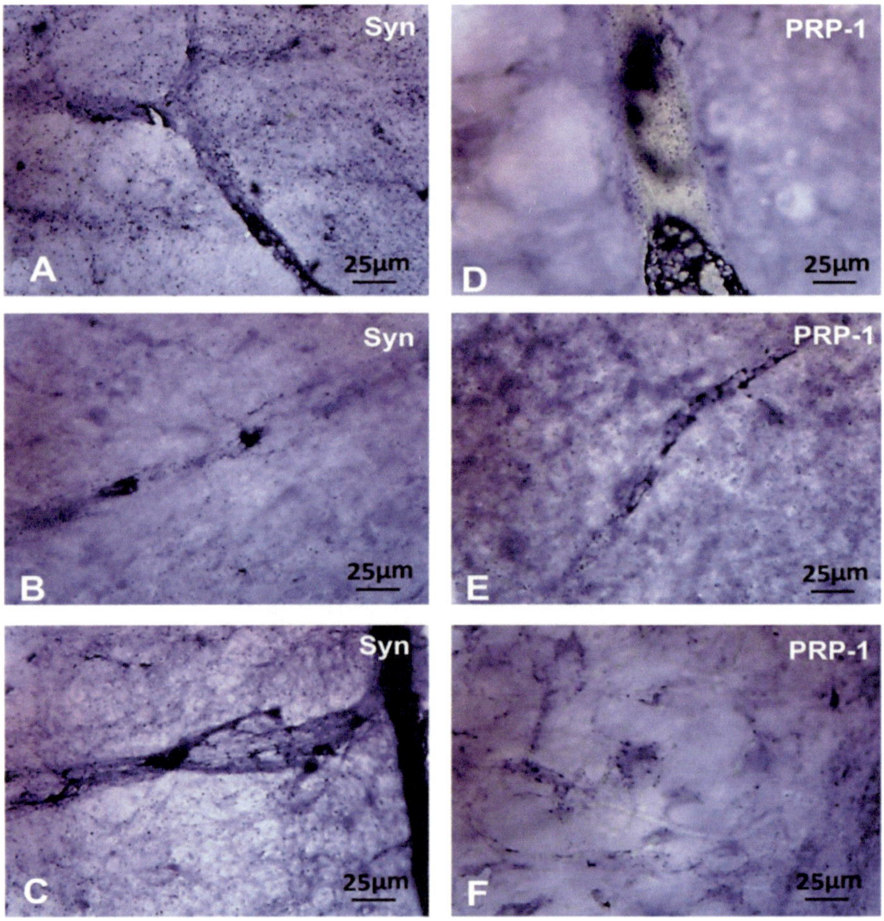

Figure 9.
Synaptophysine-Ir (Syn-Ir) and PRP-1-Ir varicose nerve fibers and varicosities in the brain of 75-days aged rats exposed to prenatal IMO stress and injected with PRP-1. Syn-Ir (A-C) and PRP-1-Ir (D-F) nerve fibers and varicosities, possibly synapses, scattered in the brain, are well demonstrated. ABC immunohistochemical method.

express neuronal marker genes, or from the neural stem/progenitor cells (NPCs) in non-neurogenic regions giving rise to the neurons mediated by the local astrocyte populations.

4. Protective effect of PRP-1 on the immune system cells death

Histochemical and immunohistochemical studies were also carried out to investigate the morpho-functional states of bone marrow (BM) structures of intact rats, rats injected with PRP-1, and rats exposed to immobilization stress and to bilateral electro-stimulation of the hypothalamic paraventricular nucleus [11].

The increased number of PRP-1-Ir blood-forming cells were observed in BM stroma and sinusoidal capillaries after PRP-1-antiserum was applied (**Figure 10**).

In addition, PRP-1-Ir varicose nerve fibers and islands of PRP-1-Ir immune system cells were detected in the surrounding areas of sinusoids (**Figure 11**).

The possibility that the PRP-1 synthesis takes place in the immune cells exists due to the evidence of distinct neuropeptides biosynthesis in the lymphocytes. Based on this, we conducted in vitro experiments in intact lymphocytes isolated from the rat bone marrow. By using PRP-1-antiserum and the flow cytofluorimetric analysis, about 4–5% PRP-1 was detected in the lymphocytes in the presence of some

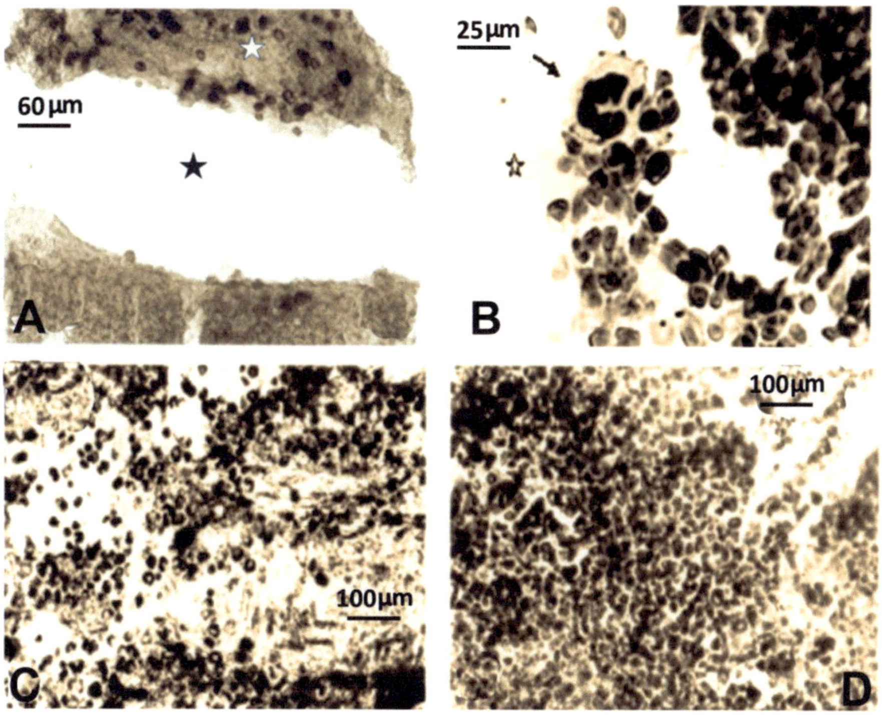

Figure 10.
Hematopoiesis in the bone marrow of rats injected with PRP-1. (A) sinusoidal capillary in the intact rat BM appeared to be empty, in general (black asterisk). Increased number of immune system cells both in sinusoid (asterisk) (B) and stroma (B-D) are well seen in the BM of injected with PRP-1 rats. (B) in the sinusoid, a megakaryocyte (arrow) with the homogenously and densely stained nuclei is demonstrated in the stage of platelets release. Histochemical method on detection of Ca2+-dependent acidic phosphatase activity.

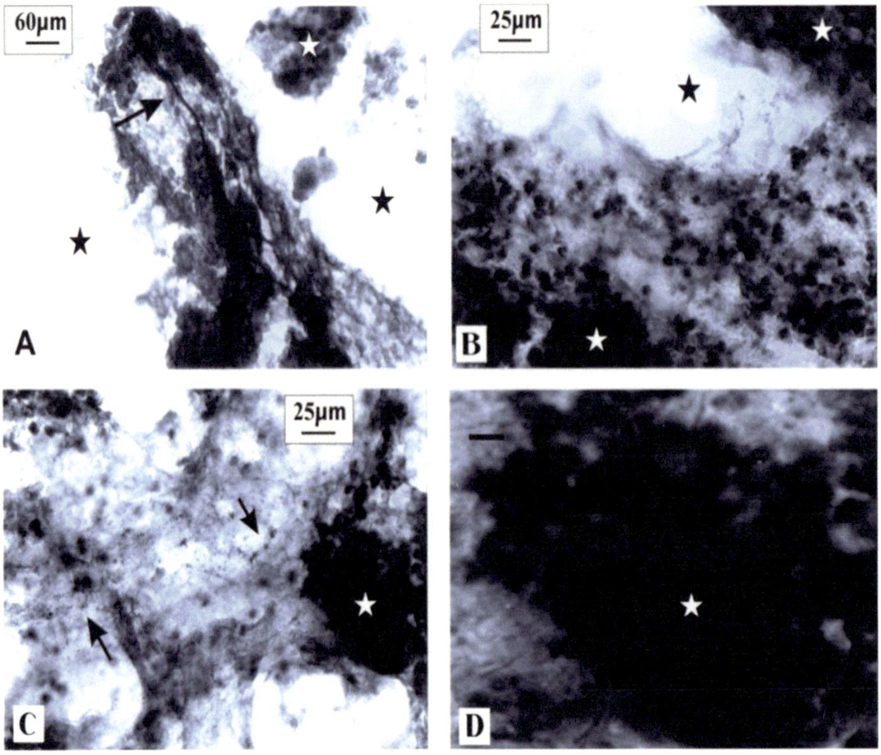

Figure 11.
PRP-1-Ir structures in bone marrow of the immobilized rats. The increased number of PRP-1-Ir blood-formed cells is detected in BM sinusoidal capillaries and stroma. Among these cells, islands of PRP-1-Ir immune system cells (white asterisks) were found inside and around the sinusoids (black asterisks). A PRP-1-Ir capillary (arrow) (A) and a few single PRP-1-Ir varicose nerve fibers (arrows) (C) are seen. (D): fragment of 10C. ABC immunohistochemical method.

activators, such as phytohemagglutinin (FGA), phorbol miristil acetate (FMA), and concanavaline A (ConA), compared to near 0% in the intact immune cells. Data obtained indicate the possible synthesis of PRP-1 in the immune system cells in vivo.

Thus, adult stem cells plasticity in response to the immobilization stress is assumed in some of the studied brain regions. However, whether these cells are indeed bone marrow-derived stem cells circulating in the blood is a question to be answered. In regards to the PRP-1-Ir migrating cells from the SC central canal, we assumed that they could be the SC stem cell-derived structures.

5. Effect of PRP-1 on inflammation

The inflammation, a physiological response to a variety of tissue damages, is made up of multiple related chains of cellular and chemical reactions. This cascade of responses activates the localized production of cytokines implicated in the cell recruitment and differentiation through specific gene expression. Though the processes behind the acute neuroinflammation following trauma or stroke may worsen the initial lesion through the increased neuronal loss, they stimulate the subsequent functional recovery through promoting neuronal plasticity. However, the consequences of chronic neuroinflammation are suspected to include neuronal loss in pathological conditions including neurodegenerative and autoimmune disorders and diseases [22].

Inflammation of the brain is linked to the biosynthesis and secretion of several neuroactive molecules, such as oxygen and nitrogen free radicals, cytokines, excitatory amino acids, proteases, complement proteins, and others, by the activated glial cells [23]. Though chronic, unregulated, and ongoing inflammation is highly prejudicial, inflammation is generally a beneficial process for the organisms due to its role in containing and curbing the survival and the spread of pathogens, as well as energy conservation and tissue recovery [24].

Beta Amyloid peptide (Aβ) (1–42) by its nature is neurotoxic and is linked to dysregulation of the brain function during Alzheimer's disease (AD) and, therefore, is strongly associated with the brain function loss through the course of AD. The accumulation of Beta Amyloid triggers the induction of neuronal cytotoxic pathways, involving microglia induced activation of pro-inflammatory cytokines IL-β and TNF-α and formation of free radicals [25, 26].

TNF-α can promote tumorigenesis leading to prostate and other cancers, and initiate apoptotic cell death [27]. Inflammatory processes are correlated with the neuronal apoptosis found in neurodegenerative diseases. Whether apoptosis plays an overall beneficial or detrimental role in neuroinflammation is unclear and topic remains controversial.

PRP-1 has roles as a caspases-2 and -6 activator [28], immunocompetent cells (T and B lymphocytes and macrophages) stimulator [8, 9], and pro-apoptotic caspases-3 and -9 inhibitor. Furthermore, it is a tumor necrosis factor alfa (TNFα) and interleukins (IL-1, IL-6) inductor in lymphocytes, astrocytes, macrophages, and fibroblasts.

PRP-1 exhibits regulatory effect on myelo- and lymphopoiesis [29–31] and neuroprotection, countering multiple toxic endogen agents [32, 33], as well as demonstrates strong neurotrophic effect on glial fibrillary acidic protein (GFAP) biosynthesis in astrocytes [34].

PRP-1's influence on staurosporine-promoted apoptosis of postnatal hippocampal cells as well as on doxorubicin-induced bone marrow mono/granulocyte apoptosis was investigated [35]. We characterized PRP-1's activity on the neuron survival rate (in a myelopoiesis context) by demonstrating the significant reduction of staurosporine-induced apoptosis of postnatal hippocampal cells from PRP-1 treatment. PRP-1's protective function against apoptosis was shown to be both dose- and time- dependent. Prolonged PRP-1 treatments showed more pronounced neuroprotection against staurosporine-induced apoptosis. A similar significant reduction was seen in bone marrow monocyte and granulocyte apoptosis by doxorubicin. The neuroprotective effect lasted for 2–4 hours and was no longer effective at 24 h when doxorubicin and PRP-1 were simultaneously added. In conclusion, the endogenous peptide PRP-1 has the primary functions of regulating myelopoiesis and neuroprotection.

Multiple experiments were carried out to understand the effect of PRP-1 on the proliferation and the colony formation of multipotent mesenchymal stromal cells (MMSCs). The dose response effect of PRP administration to rats was observed with the increased number of MMSCs in bone marrow and spleen. In ex vivo condition, the addition of PRP into the culture medium led to up to 2.5-fold increase by stimulation.

On the contrary, the proliferation was inhibited 1.5 to 2-fold in the cultures of giant-cell tumor (GCT) stromal cells, when the same PRP concentrations and cultivation periods were used. PRP-1 demonstrated also the opposite effects on the proliferation of the human bone marrow stromal cells obtained from normal humans, and the stromal cells isolated from human GCT [36].

6. Effect of PRP-1 on cell death in the cancer diseases

Other results demonstrating antitumor activity of PRP1 followed, opening up new perspective of PRP-1 antitumorigenic activity. PRP-1 induced the decay of tumor cells L929 and decreased the mitotic activity of transformed mouse fibroblast cells [37–40]. The peptide caused shrinkage of tumor in sarcoma C45 after subcutaneous injection [39].

These experimental results served as predecessors of intensive studies on other musculoskeletal malignancies, particularly chondrosarcoma. Chondrosarcoma is the second most common bone malignancy, which primarily affects the cartilage cells of the femur (thighbone), arm, pelvis, knee, and spine and even larynx, head, and neck. Chondrosarcoma is a rare disease and it does not appear to respond to either chemotherapy or radiation.

Surgical resection is the only option for the treatment, although metastatic spread to lungs occurs, eventually leading to dismal prognosis as these tumors are highly aggressive. Therefore, the search for new therapies is extremely important and urgent.

To date, only a few drugs have been identified that have been successfully shown to have clinical efficacy through the inactivation of a specific oncogene phenotype [41]. Inactivation of a single oncogene can induce cancer cells to differentiate into cells with a normal phenotype, or to undergo cellular senescence and/or apoptosis [42]. Upon MYC inactivation [43], tumors variously undergo proliferative arrest, cellular differentiation, and apoptosis.

In our most early publications, we provided experimental data indicating that PRP-1 caused inactivation of cMyc oncogene in human chondrosarcoma cells, prompting us to further investigate this peptide antitumorigenic potential [44].

The cytostatic effect of PRP-1 in human chondrosarcoma JJ012 cell line was demonstrated as 80% inhibition of cell proliferation on PRP-1 treatment in comparison with the nontreated cells. Interestingly, PRP-1 did not have any effect on immortalized chondrocytes culture, which spoke to the fact that PRP-1 selectively targeted malignant sarcoma cells and not the benign cells. Caspase-3 activity was not affected and no apoptosis was detected, thus the inhibition was due to cytostatic and not cytotoxic effect [45].

The mammalian target of rapamycin (mTORC) is an intracellular serine/threonine protein kinase, which is linked to cell growth because of its important role in nutrient signaling processes. PRP-1 was revealed as mTORC1 inhibitor, as it was able to inhibit this kinase activity in statistically significant manner [46].

The fact that the concentration lower than 10 μg/ml peptide with cytostatic effect did not inhibit mTORC1 but inhibited its target cMyc prompted us to assume the possibility of PRP-1 binding to two different receptors, facilitating the antiproliferative effect.

The cytostatic, antiproliferative action of PRP-1 was also demonstrated in the triple negative breast carcinoma MDA MB 231 cell lines [47] and the cell cycle experiments pointed on obvious stall in S phase, delaying the progression to the next stage of cell cycle upon PRP-1 treatment.

Through our study, we sought to ascertain the condition of JJ012 human malignant chondrosarcoma cells' expression of intercellular junction proteins, as well as determine the effect of the antitumorigenic cytokine PRP-1 on their expression. The experimental data suggested that tumor suppressor desmosomal proteins expression in JJ012 chondrosarcoma cells is restored and H3K9 demethylase activity comprised of a pool of JMJD1 and JMJD2 is inhibited by PRP-1, suppressing the tumorigenic potential of chondrosarcoma cells [48].

Identifying the PRP-1's receptor was very important to discern the mechanism behind its action. G protein coupled receptors (GPCR) and nuclear pathway receptor assays determined that PRP-1 receptors do not belong to nuclear or orphan receptor families, and neither were they G protein coupled. We have demonstrated in our study that the interacting partners of PRP-1 binding belong to the gel forming secreted mucin MUC5B, as well as to the innate immunity pattern recognition toll-like receptors TLR1/2 and TLR6. The experimental data indicated that the aforementioned receptors had tumor suppressive function in this cellular context [49].

When examined, the microRNA expression profiles specific to tumors showed that, throughout diverse cancers, there was widespread deregulation of these molecules. MicroRNAs have been reported to have the potential to function in disease diagnostics and therapy as novel biomarkers, as well as a novel class of tumor suppressor and oncogenes. Tumor suppressors, such as miR20a, miR125b, and miR192, were significantly upregulated by mTORC-1 inhibitor PRP-1 while onco-miRNAs, miR509-3p, miR589, miR490-3p, and miR550 were downregulated in the human chondrosarcoma JJ012 cell line [50]. The fact that PRP-1 manifests itself as a powerful epigenetic regulator was confirmed with the experiments demonstrating inhibition of BAFF Chromatin remodeling complexes [51].

Experimental results indicated that among the miRNA significantly downregulated by PRP-1 treatment was miRNA 302c. miRNA 302c is a part of the embryonic human stem cell stemness regulator cluster miR302367. miR302367 is expressed in embryonic stem cells, as well as in certain tumors, but its expression is not found in normal tissue or in adult HMSCs [52]. PRP-1 had a strong effect on chondrosarcoma and multilineage induced multipotent adult cells (embryonic primitive cell type) viability by inhibiting their proliferation. However, PRP-1 did not have any cell proliferation inhibitory action on glioblastoma, because the miR-302-367 cluster in glioblastoma exhibits an opposite effect and its expression is enough to inhibit the stemness inducing properties. The antiproliferative activity of PRP-1 and its action on downregulation of miR302c has an observed correlation that explains the peptide's opposite effects on the downregulation of miR302c targets, the stemness markers Nanog, c-Myc, and polycomb protein Bmi-1 [52].

We concluded that the inhibition of H3K9 demethylase activity by PRP-1 leads to downregulation of miR302c and its targets, defining the antiproliferative role of PRP-1.

Effects of PRP-1 on a 3D chondrosarcoma tumor model in vitro (known as spheroids) and on the cancer stem cells (CSCs) that form the spheroids, was evaluated in another study [53]. Spheroid formation and colony formation assays of cell fractions (including CSCs) were used in comparing PRP-1 treated groups with the controls. The CSCs were assessed with a modified Annexin V/propidium iodide assay for early apoptosis and cell death. Western blotting confirmed mesenchymal marker expression, and the spheroid self-renewal assay demonstrated the presence of the self-renewing CSCs. The study's results determined that PRP-1 eliminates spheroid formation and independent CSC growth, indicating the PRP-1 potential to inhibit tumor formation in a murine model. Another indication of an advantageous decline in tumor stromal cells is the decrease in non-CSC bulk tumor cells. These findings lead us to conclude that PRP-1 inhibits CSC proliferation in 3D tumor models that mimic the behavior of in vivo chondrosarcoma.

Although the cytostatic effect of PRP-1 has been demonstrated in various tumors we studied, the potential of PRP-1-related apoptosis in other types of cancer has not been ruled out.

PRP-1 action is cellular and disease context dependent. Morpho-functional study on the effect of PRP-1 on a mouse Ehrlich ascites carcinoma (EAC) model was conducted [54]. The number of viable cells in the suspension was determined by the histological method of exclusion with trypan blue (diazo live dye). The percentage of dead and alive cells was calculated after 24 h of incubation in the control samples and in those treated with PRP-1 at 0.1 and 1 μg/ml concentrations. The effect of PRP-1 on the number of tumor cells incubated for 24 h and their viability led to a 44% reduction in the number of viable cells on day 11 post-inoculation, vs. the 22% inhibition of viable cells after PRP-1 treatment (0.1 μg/ml) on day 7 post-inoculation (**Figures 12** and **13**).

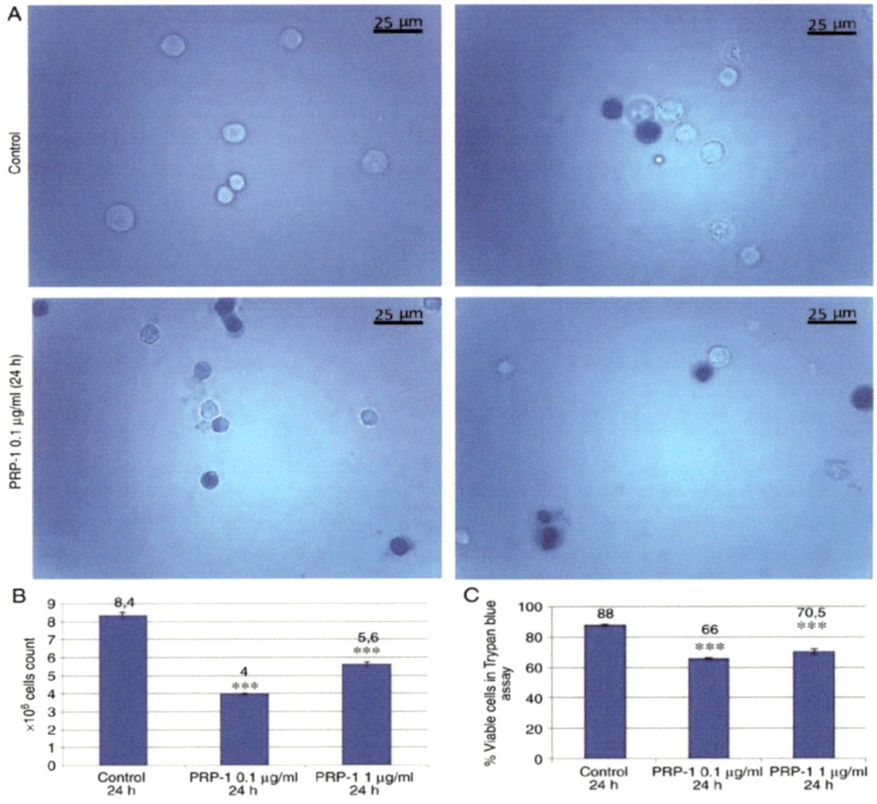

Figure 12.
Effect of the hypothalamic PRP-1 on the growth and viability of mouse isolated EAC cells on the 7th *day of tumor growth. By the histological method with Tr-Bl staining, viable EAC cells were revealed in the control samples before their culture (control). Few number of dead Tr-Bl-positive tumor cells were detected among the viable cells in the non-treated control samples 24 h after culture (control 24 h). An increased number of Tr-Bl-positive dead cells was evident 24 h after 0.1 and 1 μg/ml PRP-1 administration. The PRP-1 (0.1 and 1 μg/ml) inhibitory effect on the number of (B) total and (C) viable tumor cells treated for 24 h was statistically (***P<0.001) significant, difference compared to the control at 24 h. Histological method with Tr-Bl staining.*

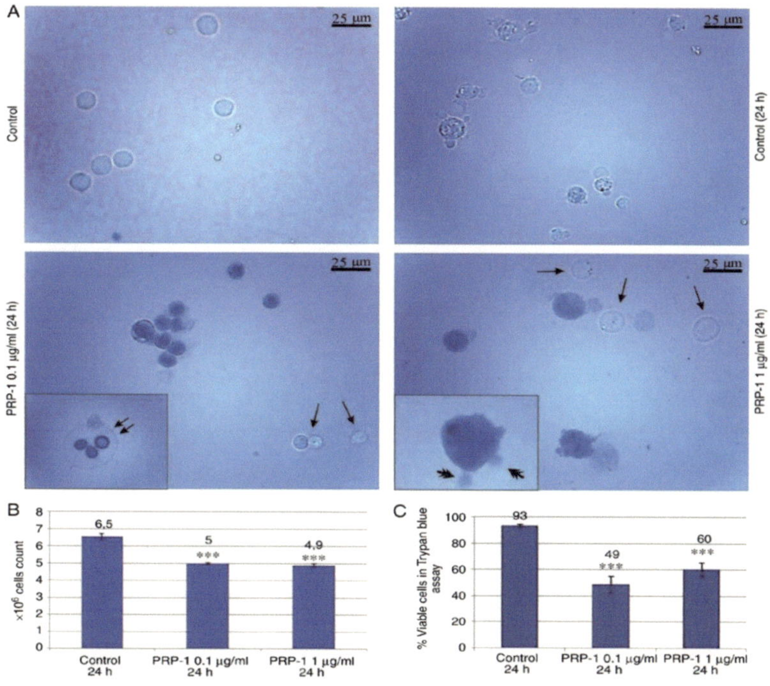

Figure 13.
*Effect of the hypothalamic PRP-1 on the growth and viability of mouse isolated EAC cells on the 11th day of tumor growth. (A) Histological method with Tr-Bl staining detected viable EAC cells before their culture (control). EAC control cells after 24 h of incubation were mainly viable, although several dead Tr-Bl-positive cells were present (control 24 h). In samples treated with 0.1 and 1 µg/ml PRP-1, an increased number of Tr-Bl-positive non-viable cells was detected; along with various viable cells (arrows), apoptotic cells with fragmented nuclei (double arrows), as well as various Tr-Bl-positive cells surrounded by apoptotic bodies (arrow heads) were observed. The PRP-1 (0.1 and 1 µg/ml) inhibitory effect on the number of (B) total and (C) viable tumor cells treated for 24 h was statistically (***P<0.001) significant, difference compared to the control at 24 h. Histological method with TR-Bl staining.*

Based on the PRP-1-induced morphological features of EAC cells, the apoptotic nature of PRP-1 was confirmed histologically as manifested by cell shrinkage, membrane blebbing, chromosome condensation (pyknosis), and nuclear fragmentation (karyorrhexis) (**Figures 14** and **15**).

To verify this observation, a series of experiments were performed, which were focused on the determination of apoptosis in cultured tumor cells using an Annexin V-Cy3 apoptosis detection kit and fluorescence microscopy. The analysis of the apoptosis on the 7-day inoculated mice EAC-cultured cells exposed to 0.1 µg/ml PRP-1 for 24 h revealed a significant increase in the number of apoptotic cells, reaching 50.33%, compared to 8.33% in the control sample on day 7. Besides, early apoptotic cells, as well as late apoptotic cells, containing and surrounded by fragments of necrotic nuclei were also detected, in contrast to the numerous viable tumor cells detected in the untreated control samples.

In late apoptotic cells, the apoptotic bodies undergo secondary necrotic changes and turn to detritus, known as a secondary form of necrosis mainly *in vitro* experiments when phagocytosis does not occur due to the absence of macrophages [55] (**Figure 16**).

In addition, a series of experiments aimed at elucidating the possible participation of PRP-1 in antitumorigenic processes was carried out by detecting the immunohistochemical localization of PRP-1 in the control and experimental EAC cells (**Figures 17** and **18**).

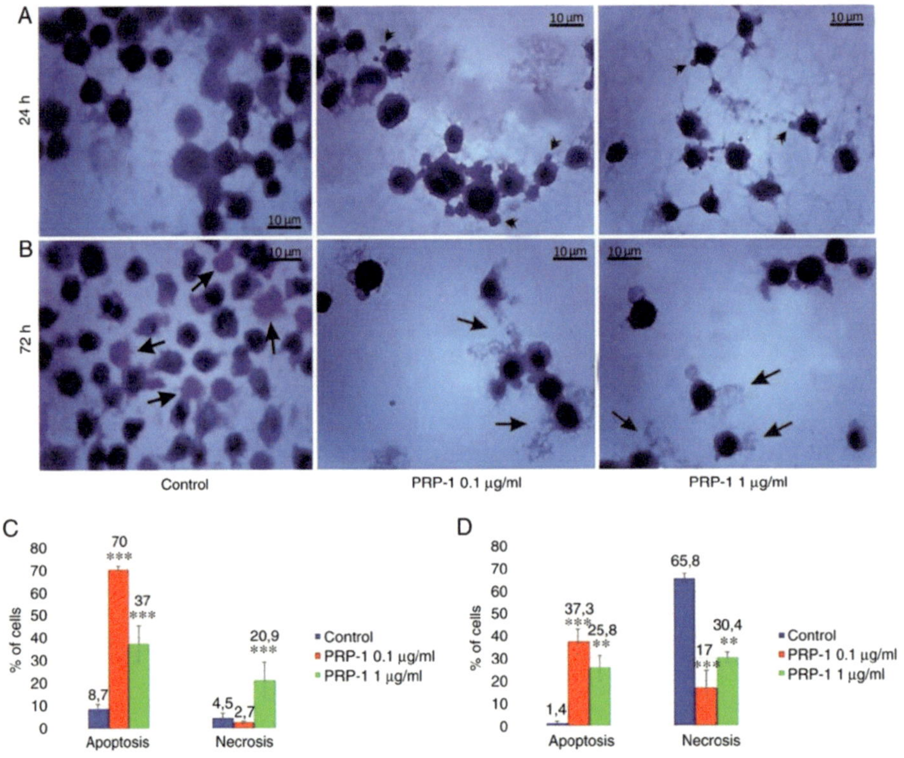

Figure 14.
*Histological evaluation of the hypothalamic PRP-1 effect on mouse-isolated EAC cells on day 7 of tumor growth. Morphological changes of tumor cells (A) 24 and (B) 72 h after culture. (A) In control samples, numerous EAC cells linked with each other were detected at 24 h, whereas a decreased number of cells was observed with both doses of PRP-1. In the experimental samples, PRP-1-induced morphological changes were similar for both the two time-points of culture. Apoptotic membrane blebbing and apoptotic bodies (arrowheads), smaller and round-shaped cells with eosinophilic cytoplasm and condensed nuclei (pyknosis), and loss of reticular extensions and contacts with adjacent cells were observed. (B) Necrotic EAC cells containing no nuclei (karyolysis) or cells with lost membrane integrity (arrows) were mainly presented in the control samples after 72 h of culture, whereas few necrotic cells with lost plasma membrane integrity and released cell death products (arrows) were detected in the samples treated with PRP-1 for 72 h. Statistical data regarding the PRP-1 (0.1 and 1 µg/ml) effect on the apoptosis and necrosis in tumor cells treated for (C) 24 h and (D) 72 h were presented according to the H&E exclusion test in comparison with the findings in the untreated control cells. Data are presented as the mean ± standard deviation (n = 3), and represent ≥3 independent experiments. **P<0.01; ***P<0.001, significant difference compared to (C) the control at 24 h and (D) the control at 72 h. Histological method with H&E staining.*

In the mice of control non-cultured EAC cells, PRP-1-IR was not detected on the seventh day of tumor growth. No intracellular PRP-1-IR was detected in the untreated control EAC cells cultured for 24 h however the presence of PRP-1-Ir cell membrane (21%) in the shape of narrow ring was registered. In the samples with 0.1 µg/ml PRP-1 treatment, strong PRP-1-IR was observed in the cytoplasm (46.5%) and nucleoli (10%).

However, the number of EAC cells with cytoplasmic PRP-1-IR of inoculated for 11 days mice and cultured for 72 h in the presence of PRP-1, constituted 73.4%, in contrast to 32.6% of the control cells with cytoplasmic PRP-1-IR. Notably, dense PRP-1-IR was noticed in 25% of apoptotic cells with membrane blebbing, in contrast to 4% of the untreated control cells.

Fluorescent nuclear staining of DAPI nevertheless proved statistically insignificant nuclear localization of PRP-1 in both control and PRP-1-treated samples after 72 h of incubation (4% of the total number of tumor cells).

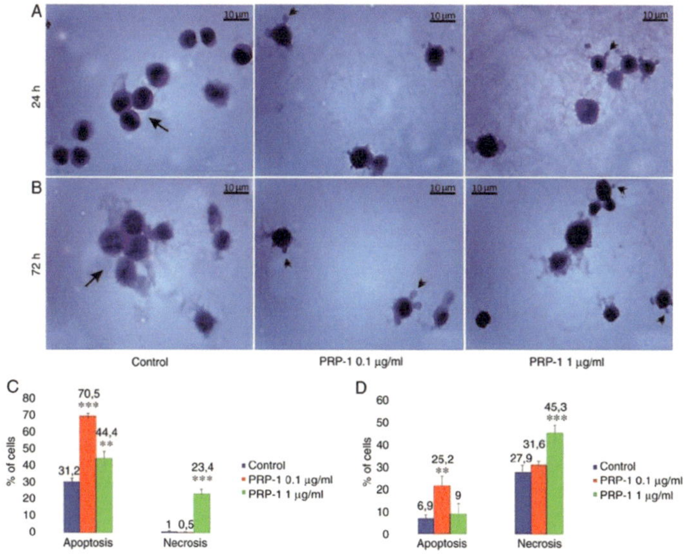

Figure 15.
*Histological evaluation of the hypothalamic PRP-1 effect on mouse-isolated EAC cells on the 11th day of tumor growth by H&E staining. Morphological changes of tumor cells at (A) 24 and (B) 72 h after culture. In the untreated control samples, tumor cells (arrows) with typical morphology were observed after both culture time-points. In comparison to cells in the control group, cells exposed to 0.1 and 1 µg/ml PRP-1 were smaller in size and exhibited a round shape. EAC cells with the apoptotic bodies (arrowheads) were observed having no contact to adjacent cells. Statistical data regarding the PRP-1 (0.1 and 1 µg/ml) effect on the apoptosis and necrosis in tumor cells treated for (C) 24 h and (D) 72 h was presented according to the H&E exclusion test in comparison with the findings in untreated control cells. Data are presented as the mean ± standard deviation (n=3), and represent ≥3 independent experiments. **P<0.01; ***P<0.001, significant difference compared to (C) the control at 24 h and (D) the control at 72 h. Histological method with H&E staining.*

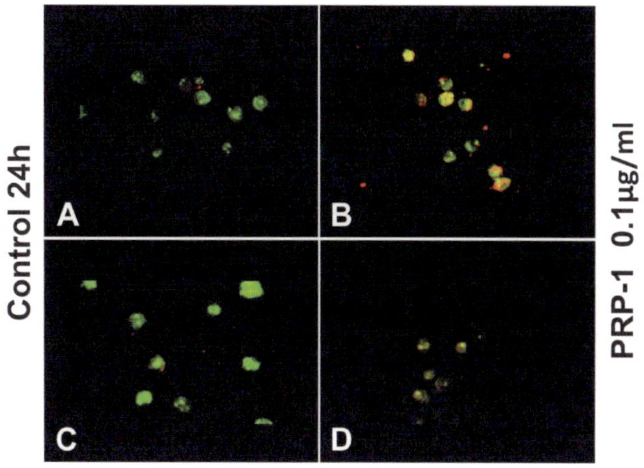

Figure 16.
Analysis of apoptosis/necrosis in cultured mouse EAC cells exposed to the hypothalamic PRP-1 on the 7th (A, B) and 11th days (C, D) of tumor growth according to fluorescence detection with Annexin V-cyanine 3. (A, C) viable EAC cultured cells (green) were detected 24 h after growing in the control untreated samples. In contrast to control samples, on the 7th day of tumor growth (B) an increased number of early apoptotic cells (orange) was revealed 24 h after treatment with 0.1 µg/ml PRP-1. Fragments of necrotic nuclei (red) were clearly detected in late apoptotic cells. (D) on the 11th day of tumor growth after treatment with 0.1 µg/ml PRP 1, the plasma membrane and certain weakly stained intracellular components could be indicative of early-stage apoptosis. Fluorescent method with Annexin V-Cy3 staining.

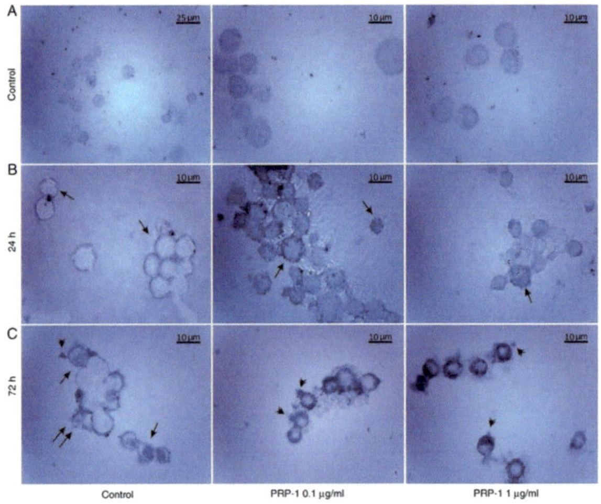

Figure 17.
Immunohistochemical localization of the hypothalamic PRP-1 in cultured mouse EAC cells on the 7th day of tumor growth. (A) All microimages demonstrated no PRP-1-IR in EAC cells before culture (control). PRP-1-IR in tumor cells (B) at 24 h and (C) 72 h after culture. (B) No intracellular PRP-1-IR was detected in control EAC cells after 24 h of culture, whereas the plasma membrane exhibited weak PRP-1-IR in the form of a narrow ring (arrows). In experimental samples, the sub-membrane cytoplasm with dense PRP-1-IR was detected in the tumor cells (arrows) exposed to 0.1 µg/ml PRP-1. (C) After 72 h of culture, nuclear localization of PRP-1 was detected in certain control (arrows) and PRP-1 treated (not shown) tumor cells. PRP-1-Ir cytoplasm was released from necrotic control cells with lost membrane integrity (double arrows). The strong PRP-1-Ir cytoplasm was revealed both in control (not demonstrated) and exposed to PRP-1 EAC cells. Notably, the apoptotic cells with the apoptotic bodies (arrowheads) also demonstrated strong PRP-1-Ir cytoplasm. ABC immunohistochemical method.

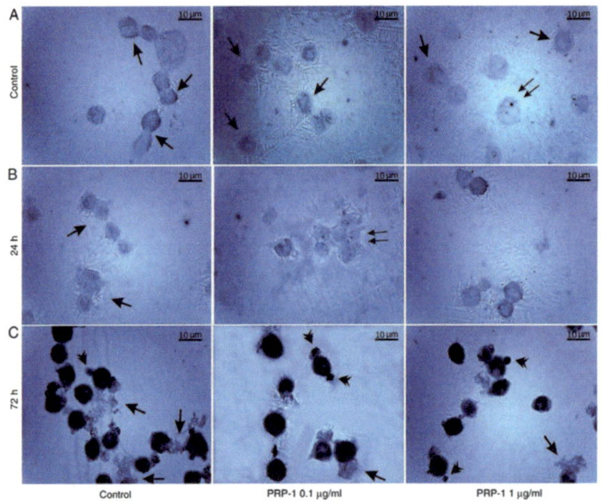

Figure 18.
Immunohistochemical localization of the hypothalamic PRP-1 in cultured mouse EAC cells on the 11th day of tumor growth according to the ABC immunohisto-chemical method. (A) EAC control cells before culture, where PRP-1 was localized in the cell membrane, cytoplasm (arrows) and nucleoli (double arrows) of certain tumor cells. PRP-1-IR in the tumor cells (B) 24 h and (C) 72 h after their culture. (B) In the untreated control samples at 24 h, weak PRP-1-IR was mainly observed in the perinuclear zone of cell cytoplasm. In the experimental samples exposed to 0.1 µg/ml PRP-1 for 24 h, PRP-1-IR was observed in the cell nucleoli (double arrows). (C) At 72 h after EAC cell culture, dense cytoplasmatic IR for PRP-1 was detected in tumor cells both in the control and PRP-1 treated samples. Morphological changes of the cells undergoing death-related processes (apoptosis and necrosis) were clearly observed, including release of PRP-1-Ir intracellular contents from necrotic cells into the extracellular space, which was detected predominantly in the control samples (arrows), while PRP-1-Ir plasma blebs and apoptotic bodies (arrowheads) were revealed mainly in the experimental samples. ABC immunohistochemical method.

Thus, the findings provide evidence that the effect of PRP-1 is cellular context dependent in EAC cells, with PRP-1 acting as a cytotoxic agent by inducing programmed cell death type I apoptosis. With regard to the detection of PRP-1-IR in the nucleus and cytoplasm of apoptotic EAC cells cultured for 72 h in both control (untreated) and experimental (PRP-1 treated) samples, the possible biosynthesis of the endogenous PRP-1 in the studied cancer cells should be taken into account.

7. PRP-1 as a circulating biomarker

Today, much attention is paid to the discovery of circulating biomarkers in the blood serum of patients with different disorders.

Previously, we succeeded in the detection and quantification of PRP-1 in rat blood serum by an enzyme linked immunosorbent assay (ELISA) developed for PRP-1 [56]. The minimum detectable concentration of PRP-1 in the intact rat blood serum has been shown to be approximately 1.78 ng/ml. Furthermore, the effect of the exogenous PRP-1 on the endogenous PRP-1 concentration was identified in the blood after 5 h and 2 days of its administration.

The significant increase of PRP-1 concentration observed in the blood samples in 5 h after the PRP-1 intraperitoneal injection was decreased in the 2-day post-injection period to approximately the control level.

Based on the recent data [57] pointing to PRP-1 being a new natural substrate for the multifunctional dipeptidyl protease (DPP-IV) that hydrolyses the peptide bonds formed by the proline residues, the decrease in the peptide concentration could be explained by the proteolytic processing of PRP-1 by DPP-IV.

The results serve as a basis for suggesting the involvement of different factors (neuropeptides, enzymes, neurotransmitters, etc.) in the mechanism of the PRP-1 action, and justify the need for additional studies for demonstrating the potential role of PRP-1 in the stress-induced disorders obtained on the animal models and in the pathogenesis of various human diseases.

Author details

Silva Abrahamyan[1*] and Karina Galoian[2]

1 NAS RA Laboratory of Histochemistry and Functional Morphology, H.Buniatian Institute of Biochemistry, Yerevan, Armenia

2 Department of Orthopedics, University of Miami, Miller School of Medicine, Miami, Florida, USA

*Address all correspondence to: silva.abrahamyan@gmail.com

IntechOpen

© 2022 The Author(s). Licensee IntechOpen. This chapter is distributed under the terms of the Creative Commons Attribution License (http://creativecommons.org/licenses/by/3.0), which permits unrestricted use, distribution, and reproduction in any medium, provided the original work is properly cited.

References

[1] Galoyan A. New polypeptides of hypothalamus: Isolation and primary structure. Proceeding NAS RA. 1996;**96**: 555-570. DOI: 10.54503/0321-1339

[2] Galoyan AA. Biochemistry of Novel Cardioactive Hormones and Immunomodulators of the Functional System Neurosecretory Hypothalamus–Endocrine Heart. Design, Nauka Publishers: Moscow; 1997. p. 240 ISBN: 5-02-022509-6

[3] Galoyan A, Sarkissian J, Kipriyan T, Sarkissian E, Sulkhanyan R, Grigorian Y, et al. Comparison of the protection against neuronal injury by hypothalamic peptides and by dexamethasone. Neurochemical Research. 2000;**25**(12):1567-1578. DOI: 10.1023/A:1026662318816

[4] Galoyan A, Sarkissian J, Kipriyan T, Sarkissian E, Chavushyan E, Sulkhanyan R, et al. Protective effects of proline-rich peptides against cobra venom and trauma induced neuronal injury. Neurochemical Research. 2001;**26**:1023-1038. DOI: 10.1023/A:1012353005489

[5] Abrahamyan S, Meliksetyan I, Galoyan A. Immunohistochemical analysis of the brain immunomodulators (proline rich polypeptide and immunophilin) in the normal and pathological conditions. In: Proceedings in the International Conference "Biochemical and Mollecular-Biological Aspects of the Brain Immune System"; Yerevan. Tsakhkadzor. RA. September 15-19 2001. pp. 62-73

[6] Abrahamyan S, Sarkissian J, Meliksetyan I, Galoyan A. Survival of trauma-injured neurons in rat brain by treatment with proline-rich peptide (PRP-1): An immunohistochemical study. Neurochemical Research. 2004;**29**(4):695-708. DOI: 10.1023/B:NERE.0000018840.19073.0b

[7] Abrahamyan S, Meliksetyan I, Chavushyan V, Aloyan M, Sarkissian J. Protective action of snake venom Naja Naja Oxiana at spinal cord hemisection. Clinical Neuroscience. 2007;**60**(3-4): 148-153

[8] Aprikyan V, Galoyan A. Antibacterial activity of a new hypothalamic peptide. Proceeding NAS RA. 1999;**99**:367-371

[9] Galoyan A, Aprikyan V. A new hypothalamic polypeptide is a regulator of myelopoiesis. Neurochemical Research. 2002;**27**(4):305-312. DOI: 10.1023/A:1014959212651

[10] Abrahamyan S, Sahakyan I, Meliksetyan I, Tumasyan N, Badalyan B, Galoyan A. Histochemical and immunohistochemical study of morpho-functional state of brain and bone marrow cell structures of rats under stress. The New Armenian Medical Journal. 2011;**5**(2):60-68

[11] Tumasyan N, Sahakyan I, Kocharyan N, Abrahamyan S. Effect of PRP-1 on the morpho-functional state of brain cellular structures of rats with various neurodegeneration. Journal "Asymmetria". 2018;**12**(4):509-518 (In russian)

[12] Palkovits M, Deli M, Gallatz K, Tóth Z, Buzás E, Falus A. Highly activated c-fos expression in specific brain regions (ependyma, circumventricular organs, choroid plexus) of histidine decarboxylase deficient mice in response to formalin-induced acute pain. Neuropharmacology.

2007;**53**(1):101-112. DOI: 10.1016/j.neuropharm.2007.04.001

[13] Zhou H, Gao J, Lu Z, Lu L, Dai W, Xu M. Role of c-Fos/JunD in protecting stress-induced cell death. Cell Proliferation. 2007;**40**(3):431-444. DOI: 10.1111/j.1365-2184.2007.00444.x

[14] Abrahamyan S, Sahakyan I, Meliksetyan I, Tumasyan N, Badalyan B, Galoyan A. Histochemical study of the brain plasticity in the condition of the vestibular system neuronal disorder and the effect of hypothalamic proline-rich-polypeptide (PRP-1). Materials of the Russian Federation Conference with International Participation dedicated to "Functional interhemispheric asymmetry and brain plasticity"; Moscow. December 13-14 2012. pp. 223-227. (In russian)

[15] Aimone JB, Jessberger S, Gage FH. Adult neurogenesis. Journal of Scholarpedia. 2007;**2**(2):2100

[16] Holtzer H. Cell lineages, stem cells and the 'quantal' cell cycle concept. In: Stem Cells and Tissue Homeostasis. Lord BI, Potten CS, Cole RJ. editor. Cambridge, New York: Cambridge University Press; 1978. pp. 1-28. ISBN: 9780521217996

[17] Leblond C. Classification of cell populations on the basis of their proliferative behavior. National Cancer Institute Monograph. 1964;**14**:119-150. PMID: 14147128

[18] Jiang Y, Jahagirdar B, Reinhardt R, Schwartz R, Keene C, Ortiz-Gonzalez X, et al. Pluripotency of mesenchimal stem cells derived from adult marrow. Nature. 2008;**418**:41-49. DOI: 10.1038/nature00870

[19] Abrahamyan S, Meliksetyan I, Sahakyan I, Tumasyan N, Badalyan B, Galoyan A. Brain plasticity of exposed to prenatal immobilization stress rats. Biopolymers and Cell. 2011;**27**(5):339-343

[20] Lu P, Blesch A, Tuszynsky M. Induction of bone marrow stromal cells to neurons: Differentiation, transdifferentiation, or artefact? Journal of Neuroscience Research. 2004;**77**:174-191. DOI: 10.1002/jnr.20148

[21] Vogel W, Grunebach F, Messam C, Kanz L, Bruger W, Buhring H. Heterogeneity among human bone-marrow-derived mesenchymal stem cells and neural progenitor cells. Haematologica. 2003;**88**:126-133 PMID: 12604402

[22] Chauveau F, Boutin H, Van Camp N, Dollé F, Tavitian B. Nuclear imaging of neuroinflammation: A comprehensive review of [11C] PK11195 challengers. European Journal of Nuclear Medicine and Molecular Imaging. 2008;**35**:2304-2319. DOI: 10.1007/s00259-008-0908-9

[23] Gahtan E, Overmier B. Inflammatory pathogenesis in Alzheimer's disease: Biological mechanisms and cognitive sequel. Neuroscience & Biobehavioral Reviews. 1999;**23**:615-633. DOI: 10.1016/s0149-7634(98)00058-x

[24] Allan S, Rothwell N. Inflammation in central nervous system injury. Philosophical Transactions of the Royal Society of London. Series B, Biological Sciences. 2003;**358**:1669-1677. DOI: 10.1098/rstb.2003.1358

[25] Monsonego A, Imitola J, Zota V, Oida T, Weiner LL. Microglia-mediated nitric oxide cytotoxicity of T cells following amyloid β-peptide presentation to Th1 cells. Journal of Immunology. 2003;**171**:2216-2224. DOI: 10.4049/jimmunol.171.5.2216

[26] Royuela M, Rodriguete-Berriguete G, Fraile B, Paniagua R. TNF-α/IL-1/NF-κB

transduction pathway in human cancer prostate. Histology and Histopathology. 2008;**23**:1279-1290. DOI: 10.14670/HH-23.1279

[27] Abdi A, Sadraie H, Dargahi L, Khalaj L, Ahmadiani A. Apoptosis inhibition can be threatening in Aβ-induced neuroinflammation, through promoting cell proliferation. Neurochemical Research. 2011;**36**:39-48. DOI: 10.1007/s11064-010-0259-3

[28] Galoyan A, Terio T, Berg M, Marks N. Effects of prolin-rich peptide derived from Neurophysin-II on caspases of murine neuroblastoma: Evidences for caspase-2 and -6 activation. Neurochemistry (RAS and NAS RA). 2000;**17**:185-188. DOI: 10.1007/s11064-005-2686-0

[29] Galoian K, Luo S, Qureshi A, Patel P, Price R, Morse A, et al. Proline rich polypeptide-1 and inflammatory pathway signaling in chondrosarcoma. Molecular and Clinical Oncology. 2016;**5**:618-624. DOI: 10.3892/mco.2016.1010

[30] Durgaryan A, Matevosyan M, Seferyan T, Sargsyan M, Grigoryan S, Galoian K, et al. The protective and immunomodulatory effects of hypothalamic proline-rich polypeptide galarmin against methicillin-resistant Staphylococcus aureus infection in mice. European Journal of Clinical Microbiology & Infectious Diseases. 2012;**3**:2153-2165. DOI: 10.1007/s10096-012-1550-9

[31] Galoyan A, Korochkin L, Rybalkina E, Pavlova G, Saburina I, Zaraiski E, et al. Hypothalamic proline-rich polypeptide enhances bone marrow colony-forming cell proliferation and stromal progenitor cell differentiation. Cell Transplantation. 2008;**17**:1061-1066. DOI: 10.3727/096368908786991579

[32] Galoyan K, Scully S, Galoyan A. Myc-oncogene inactivating effect by proline rich polypeptide (PRP-1) in chondrosarcoma. Neurochemical Research. 2009;**34**:379-385. DOI: 10.1007/s11064-008-9794-6

[33] Galoian K, Temple T, Galoyan A. Cytostatic effect of the hypothalamic cytokine PRP-1 is mediated by its inhibition of mTOR and cMyc in high grade metastatic chondrosarcoma. Neurochemical Research. 2011;**36**:812-818. DOI: 10.1007/s11064-011-0406-5

[34] Chekhonin V, Gurina O, Ruabukin I, Savchenko E, Galoyan A. Immunochemical study of the effect of neurotrophins on the GFAP synthesis in astrocyte culture. 2001. In: Proceed. of the Intern. Conf. on Bioch. and Molec. Biol. Aspects of the Brain Immune System; Yerevan-Tsakhkadzor. 15-19 September. 2001. pp. 140-145

[35] Galoyan A, Josef K, Klumpp S, Danielian K, Galoian K, Kremer W, et al. Effect of hypothalamic proline-rich peptide (PRP-1) on neuronal and bone marrow cell apoptosis. Neurochemical Research. 2007;**32**:898-1905. DOI: 10.1007/s11064-007-9379-9

[36] Chailakhyan R, Gerasimov Y, Chailakhyan M. Proline-rich hypothalamic polypeptide has opposite effects on the proliferation of human normal bone marrow stromal cells and human giant-cell tumour stromal cells. Neurochemical Research. 2010;**35**:934-939. DOI: 10.1007/s11064-009-0113-7

[37] Galoyan A, Shakhlamov V, Kondakova L, Altukhova V, Polyakova G. The influence of proline-rich polypeptide on the morphology and mitotic activity of neurinoma Hassel node of tumor cells at rats (electron microscopic investigations). PNAS (Russian Academy of Sciences). 2001;**101**(2):279-286

[38] Galoyan A, Shakhlamov V, Malaytsev V. Changes in tumor cells L_{929} under PRP effect in vitro. Medical Science of Armenia. 2001;**41**(1):25-29

[39] Simonyan G, Nersesyan A, Simonyan R, Babayan M, Simonyan M, Galoyan A. Antitumor and antistressor effects of hypothalamic proline rich peptide (PRP-1) in sarcoma 45; possible biochemical mechanisms. Neurochemistry (RAS and NAS RA). 2005;**22**(2):125-130

[40] Galoian K, Scully S, Galoyan A. Myc-oncogene inactivating effect by proline rich polypeptide (PRP-1) in chondrosarcoma JJ012 Cells. Neurochemical Research. 2009;**34**(2):379-385. DOI: 10.1007/s11064-008-9794-6

[41] Driker B, Talpaz M, Resta D, Peng B, Buchdunger E, Ford J, et al. Efficacy and safety of a specific inhibitor of the BCR-ABL tyrosine kinase in chronic myeloid leukemia. The New England Journal of Medicine. 2001;**344**:1031-1037. DOI: 10.1056/NEJM200104053441401

[42] Felsher D. Oncogene addiction versus oncogene amnesia: Perhaps more than just a bad habit? Cancer Research. 2008;**68**(9):3081-3086. DOI: 10.1158/0008-5472.CAN-07-5832

[43] Shachaf C, Felsher D. Tumor dormancy and Myc inactivation: Pushing cancer to the brink of normalcy. Cancer Research. 2005;**65**:4471-4474. DOI: 10.1158/0008-5472.CAN-05-1172

[44] Galoian K, Scully S, Galoyan A. Myc-oncogene inactivating effect of proline rich polypeptides (PRP-1) in the human chondrosarcoma JJ012 cells. Neurochemical Research. 2009;**34**(2):379-385. DOI: 10.1007/s11064-008-9794-6

[45] Galoian K, Scully S, Galoyan A. Antitumorigenic effect of brain proline rich polypeptide-1 in human chondrosarcoma. Neurochemical Research. 2009;**34**(12):2117-2121. DOI: 10.1007/s11064-009-0009-6

[46] Galoian K, Temple T, Galoyan A. Cytostatic effect of the hypothalamic cytokine PRP-1 is mediated by mTOR and cMyc inhibition in high grade chondrosarcoma. Neurochemical Research. 2011;**36**(5):812-818. DOI: 10.1007/s11064-011-0406-5

[47] Galoian K, Temple T, Galoyan A. Cytostatic effect of novel mTOR inhibitor, PRP-1 (galarmin) in MDA 231 (ER-) breast carcinoma cell line. PRP-1 inhibits mesenchymal tumors. Tumour Biology. 2011;**32**:745-751. DOI: 10.1007/s13277-011-0176-3

[48] Galoian K, Qureshi A, Wideroff G, Temple H. Restoration of desmosomal junction protein expression and inhibition of H3K9-specific histone demethylase activity by cytostatic proline-rich polypeptide-1 leads to suppression of tumorigenic potential in human chondrosarcoma cells. Molecular and Clinical Oncology. 2015;**3**(1):171-178. DOI: 10.3892/mco.2014.445

[49] Galoian K, Abrahamyan S, Chailyan G, Qureshi A, Patel P, Metzer G, et al. Toll like receptors TLR1/2, TLR 6 and MUC5b as binding interaction partners with cytostatic proline rich polypeptide 1 in human chondrosarcoma. International Journal of Oncology. 2018;**52**(1):139-154. DOI: 10.3892/ijo.2017.4199

[50] Galoian K, Guettouche T, Issac B, Qureshi A, Temple H. Regulation of onco and tumor suppressor miRNAs by mTORC1 inhibitor PRP-1 in human chondrosarcoma. Tumour Biology. 2014;**35**(3):2335-2341. DOI: 10.1007/s13277-013-1309-7

[51] Moran A, Hoyt A, Sedani A, Granger C, Blonska M, Zhao-Jun L, et al. Proline-Rich-Polypeptide (PRP-1) drastically decreases cancer stem cell (CSC) population targeting BAFF chromatin-remodeling-complexes in human chondrosarcoma JJ012-cells. Oncology Reports. 2020;**44**:393-403. DOI: 10.3892/or.2020.7612

[52] Galoian K, Qureshi A, Dippolito G, Schiller P, Molinari M, Johnstone A, et al. Epigenetic regulation of embryonic stem cell marker miR302C in human chondrosarcoma as determinant of antiproliferative activity of proline rich polypeptide 1. International Journal of Oncology. 2015;**47**:465-472. DOI: 10.3892/ijo.2015.3054

[53] Granger C, hoyt A, Moran A, Becker B, Sedani A, Saigh S, et al. Cancer stem cells as a therapeutic target in 3D tumor models of human chondrosarcoma: An encouraging future for proline rich polypeptide-1. Molecular Medicine Reports. 2020;22(5):3747-3758. DOI: 10.3892/mmr.2020.11480

[54] Abrahamyan S, Sahakyan I, Tumasyan N, Kocharyan N, Simonyan A, Aroutiounian R, et al. Morpho-functional study of the hypothalamic proline-rich polypeptide apoptotic activity against mouse Ehrlich ascites carcinoma. Oncology Reports. 2020;**44**:196-212. DOI: 10.3892/or.2020.7604

[55] Shacter E, Williams JA, Hinson RM, Senturker S, Lee Y. Oxidative stress interferes with cancer chemotherapy: Inhibition of lymphoma cell apoptosis and phagocytosis. Blood. 2020;**96**:307-313

[56] Abrahamyan S, Davtyan T, Khachatryan A, Tumasyan N, Sahakyan I, Harutyunyan H, et al. Quantification of the hypothalamic proline rich polypeptide-1 in rat blood serum. Neurochemical Journal. 2014;**36**(1):38-43. DOI: 10.1134/S1819712414010024

[57] Antonyan A, Sharoyan S, Mardanyan S, Galoyan A. Proline-rich cytokine from neurosecretory granules: A new natural substrate for dipeptidyl peptidase IV. Neurochemical Research. 2011;**36**(1):34-38. DOI: 10.1007/s11064-010-0258-4

Section 2

Ferroptosis

Chapter 3

Ferroptosis in Leukemia: Lessons and Challenges

Baoquan Song and Leisheng Zhang

Abstract

Ferroptosis is a newly defined programmed cell death (PCD) process with the hallmark of the accumulation of iron-dependent lipid peroxidation, which is more immunogenic over apoptosis. Ferroptosis shows great potential as a therapeutic target against acute kidney injury (AKI), cancers, cardiovascular diseases, neurodegenerative diseases, and hepatic diseases. Accumulating evidence has highlighted that ferroptosis plays an unneglectable role in regulating the development and progression of multiple pathologies of leukemia including acute myeloid leukemia (AML), chronic myeloid leukemia (CML), acute lymphoblastic leukemia (ALL), and chronic lymphocytic leukemia (CLL). Herein, we focus on the state-of-the-art renewal in the relationship of ferroptosis with leukemia. Meanwhile, this chapter further highlights the iron, lipid and amino acid metabolism, as well as ferroptosis-based molecular mechanisms. Collectively, we summarize the contribution of ferroptosis to the pathogenesis of leukemia and discuss ferroptosis as a novel therapeutic target for different types of leukemia.

Keywords: ferroptosis, programmed cell death, leukemia, metabolism, novel therapeutic targets, lipid peroxides

1. Introduction

Ferroptosis is a new type of iron-dependent programmed cell death (PCD) that was first reported by Dixon et al. in 2012, which is initiated by lipid peroxidation and terminates in toxic lipid peroxidation and mitochondrial dysfunction [1, 2]. As a new form of PCD, ferroptosis can be distinguished from other types of cell death such as necrosis, pyrolysis, apoptosis, and autophagy in regard to morphology and biochemistry [3, 4]. As to the morphological characteristics, the major manifestation of ferroptosis is increase in mitochondrial membrane density, unspoilt cytomembranes, cell volume shrinkage, decline or even deficiency in mitochondrial cristae, outer membrane rupture and mitochondrial membrane crumpling, and normal cellular nucleus size with unconsolidated chromatin [5, 6]. As to the biochemical characteristics, ferroptosis results from the accumulation of iron-catalyzed lipid-based reactive oxygen species (ROS), which is commonly initiated by either the loss of the activity of the lipid repair enzyme named glutathione peroxidase 4 (Gpx4) or the inactivation of the antioxidant defenses depending on the cellular glutathione (GSH) [7]. As to the genetic characteristics, a considerable number of genes have been indicated to

IntechOpen

modulate ferroptosis by acting as inhibitors or inducers, which can be divided into the system Xc-, Gpx4, GSH, lipid peroxidation-associated genes, and iron metabolism regulation-associated genes on the basis of the variations in targets, whereas the specific regulatory mechanisms of ferroptosis still need to be further explored [8].

Taken together, ferroptosis has been regarded with initiation by the failure of the aforementioned GSH-dependent antioxidant defenses, which thus results in the uncontrolled process during lipid peroxidation and the concomitant cell death [9, 10]. Similarly, the out-of-balance phenomenon between GSH-dependent Gpx4 inactivity and the iron-catalyzed lipid ROS production eventually triggers the occurrence of ferroptosis. Accordingly, the lipophilic antioxidants and iron chelators can effectively suppress the process of ferroptotic cell death [11]. The former inhibits the initiation and accumulation of the lipid peroxidation by capturing or eliminating targeting lipoxygenase and free radicals, including ferrostatin-1 (Fer-1), *N*-acetylcysteine (NAC), vitamin E, and liproxstatin-1 (Lip-1) [12]. The latter can efficaciously prevent the aforementioned iron-catalyzed-associated lipid peroxidation by acerating the depletion of free iron, including deferiprone (DFP) and deferoxamine (DFO) [13]. In spite of the rapid development of the multifaceted assumptions and considerable validations, the systematic and detailed molecular mechanisms of ferroptosis are still far from being fully clarified. The regulatory mechanisms of ferroptosis are complicated and involve a variety of metabolic networks and signaling pathways, including abnormal iron metabolism, lipid metabolism, amino acid metabolism, and signaling pathways associated with ferroptosis. For instance, a number of studies have reported the involvement of several signaling pathways with ferroptosis such as ferritinophagy, iron and amino acid metabolism, cell adhesion, and Keap1/Nrf2, mTOR, and p53 signaling pathways [14–16].

Nowadays, ferroptosis has been considered to perform a critical role in various diseases and pathologies, such as cancer, stroke, cardiomyopathy, kidney and liver injury, and neurodegeneration [17, 18]. Meanwhile, both the indicated inhibitors and activators have been continuously identified and introduced to explore the molecular mechanisms of ferroptosis, which collectively benefit the development of novel therapeutic strategies for the administration of ferroptosis-related diseases and pathologies [19]. In the meantime, accumulating evidence has indicated the changes in iron metabolism during leukemia, which thus has been considered as a crucial feature as well. To date, high oxidative stress and high iron requirements are identified with association to the alteration of iron metabolism during leukemia, which thus suggests that leukemia cells are more vulnerable to variations in ROS and iron levels when compared with the normal cells [20]. Therefore, targeting iron metabolism may provide new insights into approaches to the treatment of leukemia.

2. Ferroptosis and leukemia

Leukemia is a group of heterogeneous hematopoietic stem cell (HSC) malignancies. It is characterized by aberrant accumulation of undifferentiated blasts capable of unrestrained proliferation in the bone marrow, which interferes with the production of normal blood cells. Leukemic cells uniquely possess the innate ability for migration and invasion. Differentiated, malignant leukocytes retain the benign leukocytes' capacity for cell motility and survival in the circulation, while acquiring the potential for rapid and uncontrolled cell division. Currently, a variety of cancer cells of hematological malignancies (e.g., multiple myeloma, leukemia, and lymphoma) have been identified to be sensitive to ferroptosis because large amounts of iron are acquired by leukemia cells for the maintenance of rapid growth and proliferation. As a consequence, targeting iron metabolism holds the potential to supply new insights into developing novel approaches for the administration of leukemia.

As a cancer of the bone marrow and blood cells, leukemia threatens human health seriously. Recent insights into iron metabolism along with the recent discovery of ferroptosis have opened new avenues in the field of antitumor therapies. Emerging evidence has revealed that ferroptosis is essentially a nexus among metabolism, redox biology, and diseases including cancers. The discovery of ferroptosis is a major breakthrough in the development of cancer treatments. Therefore, targeting ferroptosis may provide new insights into approaches to the treatment of leukemia. Leukemia is classified into four main subgroups, including acute myeloid leukemia (AML), acute lymphoblastic leukemia (ALL), chronic myeloid leukemia (CML), and chronic lymphoblastic leukemia (CLL). Therefore, we focus on these entities in this chapter.

2.1 Ferroptosis in acute myeloid leukemia

AML is the most common type of leukemia in adults, which is characterized by the rapid growth of abnormal lineage-specific hematopoietic precursor cells that do not differentiate into functional granulocytes or monocytes during hematopoiesis in the bone marrow microenvironment [21–24]. As a classical paradigm of myeloid disease with multiple life-threatening complications, AML is characterized by a significant reduction of physiological differentiation of hematopoietic stem and progenitor cells (HSPCs) toward myeloid and lymphoid lineages in parallel with aberrant activation of pathological hematopoiesis that is dominated by the continuous accumulation of the dysfunctional leukemic blast populations [25, 26]. Despite the dramatic progresses in exploring the pathogenesis and exploring advanced targeted therapies, the majority of the AML patients are still bearing immune dysregulation and the concomitant outcomes [25]. Long-term survival remains limitation with the standard-of-care chemotherapies combining anthracyclines with cytarabine. Development of resistance to chemotherapeutic agents is a major hurdle in the effective treatment of patients with AML [27]. New therapies are needed to improve chemotherapy efficacy in AML [25, 28]. Meanwhile, many endeavors have been made to ascertain *de novo* biomarkers and improve risk stratification and prognostic assessment in different AML subgroups [21]. However, more and more studies have proved that ferroptosis is closely related to the pathophysiology of AML and thus shed light on studying the pathogenesis of AML and search for new therapeutic targets.

A variety of molecular and pathological changes in ferroptosis have been observed in experimental AML models and samples from AML patients (**Table 1**). Among ferroptosis-related genes (FRGs), GPX-1, GPX-3, GPX-4, and GPX-7 were highly expressed in *n* AML patient samples and associated with poorer prognosis of overall survival (OS) [30]. AKR1C2 and SOCS1 are promising biomarkers for predicting prognosis in patients with AML [31]. Huang et al. [32] established a prognostic model of 12-FRGs in AML. The model successfully divided patients into high- and low-risk (LR) patient groups with mean OS as the basis. Wei et al. [15] showed that ferroptosis-related genes (FRGs), DPP4 and TFRC, act as biomarkers for predicting and diagnosing AML, and their expression levels also have significant correlations with drug resistance in AML. Other markers of ferroptosis, among the 12 ferroptosis-related genes (PHKG2, HSD17B11, STEAP3, HRAS, ARNTL, CXCL2, SLC38A1, PGD, ENPP2, ACSL3, DDIT4, and PSAT1), were screened to generate a prognostic model,

Leukemia	FRGs	Risks	Highlights of the study	Ref.
AML	GPX-1, -3, -4, and -7	Poor prognosis	The study offered novel insights into the differential expression and prognostic potential of the GPX family in AML.	[12]
	ACSL6 and G3BP1	Favorable prognosis	FRG risk model may be beneficial to the precision immunotherapy of AML patients in the future, especially those of HR groups.	[14]
	GPX4, CD44, FH, CISD1, SESN2, LPCAT3, AIFM2, ACSL5, HSPB1, and SOCS1	Favorable prognosis		
	CHAC1, CISD1, DPP4, GPX4, AIFM2, SQLE, PGD, and ACSF2	Poor prognosis	A novel ferroptosis-related prognostic model for outcome prediction and risk stratification in AML was conducted and validated.	[15]
	ZFPM2, ZNF560, ZSCAN4, HMX2, HRASLS, LGALS1, LHX6, CCL23, and FAM155B	Poor prognosis	The identified genes were affected by ferroptosis and develop a prognostic risk-scoring model to predict patients' survival at the genetic level.	[16]
	MXRA5, PCDHB12, PRINS, TMEM56, TWIST1, ASTN1, DLL3, EFNB3, and FOXL1	Favorable prognosis		
	AP001266.2, AC007383.2, AC008906.1, AC026771.1, and KIF26B-AS1	Favorable prognosis	Seven novel ferroptosis-related lncRNA signatures were established to accurately predict the prognosis of AML.	[17]
	AC133961.1 and AF064858.3	Poor prognosis		
	AKR1C2 and SOCS1	Poor prognosis	A prognostic risk model that included AKR1C2 and SOCS1 predicted outcomes in AML patients.	[18]
	DNAJB6 and HSPB1	Poor prognosis	Potential targets and new research ideas for the treatment and early detection of AML were identified.	[19]
	HIVEP3	Poor prognosis	HIVEP3 is a *de novo* independent prognostic indicator and the crosstalk between HIVEP3 and ferroptosis signaling pathways.	[20]
	Dipetidyl peptidase-4	Poor prognosis	DPP4 as a biomarker for predicting and diagnosing AML influences drug resistance in AML.	[21]
CML	TP63, STEAP3, NQO1, and ELAVL1	Poor prognosis	Cysteine depletion serves as a potential therapeutic strategy for overcoming chemotherapy resistance in CML.	[29]
	PRKAA1, HELLS, FANCD2, and CDKN2A	Favorable prognosis		

Table 1.
Bioinformatics studies predicting prognosis of leukemia patients based on the expression of ferroptosis-related genes

which stratified patients into a low- (LR) or high-risk (HR) group [33]. Another study showed that 18 signature genes, including DLL3, EFNB3, ZSCAN4, ASTN1, FAM155B, CCL23, ZFPM2, FOXL1, HMX2, LGALS1, LHX6, PCDHB12, MXRA5, HRASLS, TMEM56, PRINS, TWIST1, and ZNF560, were unified for the development of establishing the prognostic risk-scoring model. With the aid of the model, AML patients can be grouped into high-risk and low-risk groups, and those inpatients with low risk consistently revealed preferable survival over the high-risk inpatients [34]. In another study, investigators have identified and verified seven ferroptosis-related lncRNA signatures (AP001266.2, AC133961.1, AF064858.3, AC007383.2, AC008906.1, AC026771.1, and KIF26B-AS1) with independent prognostic value in patients with AML (**Table 1**) [35]. In summary, we conducted and validated a novel ferroptosis-related prognostic model for outcome prediction and risk stratification in AML, with great potential to guide individualized treatment strategies in the future [31, 36–38].

Currently, ferroptosis has been characterized by the well-established iron-dependent accumulation of the lipid hydroperoxides, which eventually leads to the severe impairments of the mitochondrial outer membrane as well as the decrease of mitochondria crista. Interestingly, cancer cell death has been proved to be involved in ferroptosis as well. Therewith, new treatment remedies by developing effective activators and inhibitors targeting ferroptosis will benefit the development of novel treatment paradigms for AML patients. As early as 2015, researchers found that the ferroptosis inducer Erastin enhances sensitivity of acute myeloid leukemia cells to chemotherapeutic agents [39]. Later, researchers demonstrated that DHA would induce autophagy and ferroptosis in AML cell lines and revealed the role of iron metabolism in DHA-induced cell death [40]. High mobility group box 1 (HMGB1) is a novel regulator of ferroptosis *via* the RAS-JNK/p38 pathway and a potential drug target for therapeutic interventions in leukemia. It plays an important role in leukemia pathogenesis and chemotherapy resistance [29]. Typhaneoside (TYP) is a major flavonoid in the extract of pollen typhae, showing significant biological and pharmacological effects. Zhu et al. [24] found that TYP significantly triggered autophagy in AML cells by promoting the activation of AMP-activated protein kinase (AMPK) signaling, contributing to ferritin degradation, ROS accumulation, and ferroptotic cell death ultimately. APR-246, also known as PRIMA-1MET, is a promising new therapeutic agent that targets TP53-mutated cancers. The association of APR-246 with induction of ferroptosis (either by pharmacological compounds or by genetic inactivation of SLC7A11 or GPX4) had a synergistic effect on the promotion of cell death, both *in vivo* and *ex vivo* [41]. circKDM4C is negatively associated with AML, and the downregulated circKDM4C leads to AML progression, which otherwise induces ferroptosis by regulating has-let-7b-5p and P53. This may be explored further to develop a potential AML therapy [42]. Aldehyde dehydrogenase 3a2 protects AML cells from oxidative death and the synthetic lethality of ferroptosis inducers. Combination of Aldh3a2 inhibition with ferroptosis inducers or with standard AML induction chemotherapy deserves further consideration as a cancer therapy [43]. Subsequent studies verified that HMOX1 was a critical target in honokiol-induced ferroptosis [44]. These results reveal that honokiol is an effective antileukemia agent in AML cell lines and may be a potential ferroptosis activator in AML.

2.2 Ferroptosis in acute lymphoblastic leukemia

Acute lymphoblastic leukemia (ALL) is a malignant clonal disorder of lymphoblastic hematopoiesis with high heterogeneity [45, 46]. Survival rates of ALL have improved remarkably by intensive induction chemotherapy, with complete remission (CR) rates of up to 80%. However, relapse occurred in patients ranging from 25% to 35% [47]. Interestingly, ferroptosis is suggested to be a promising strategy for cancer treatment and therefore should also be evaluated in ALL [48, 49].

After the exploration of the potential role of ferroptosis in Ph-neg B-ALL with the clinical data and the RNA-seq results of 80 Ph-neg B-ALL, a Cox regression model based on 8 FRGs (ALOX15, ATP5G3, CARS, CDKN1A, LPCAT3, SAT1, SLC1A5, and TFRC) was established to help evaluate the prognosis of Ph-neg B-ALL patients [50]. Lukas et al. reported that the glutathione (GSH) peroxidase 4 (GPX4) inhibitor RSL3 triggers lipid peroxidation, production of reactive oxygen species (ROS) and cell death in ALL cells. Importantly, LOX inhibitors, including the selective 12/15-LOX inhibitor Baicalein and the pan-LOX inhibitor nordihydroguaiaretic acid (NDGA), protect ALL cells from RSL3-induced ferroptosis [51]. Artesunate (ART), a widely used antimalarial compound, exerted potent anti-ATLL effects through inducing reactive oxygen species production, resulting in cell death mediated by apoptosis, ferroptosis, and necroptosis [52]. Greco et al. [53] reported that sulforaphane (50 μM) induced U-937 cell ferroptosis through depletion of glutathione (GSH), decreased GSH peroxidase 4 protein expression, and lipid peroxidation. PAQR3 (also known as RKTG) has been proved to take part in many human cancers by acting as a tumor suppressor. Jin and Tong [48] showed PAQR3 inhibited proliferation and aggravated ferroptosis in ALL through modulation of Nrf2 stability. This study suggested that PAQR3 may serve as an effective biological marker for ALL treatment. Meanwhile, Hydnocarpin D (HD) is a bioactive flavonolignan compound that possesses promising antitumor activity, although accumulation of lipid ROS and decrease of GSH and GPX4, while inhibition of autophagy, impeded ferroptotic cell death [54]. Poricoic acid A (PAA) is the main chemical constituent on the surface layer of the mushroom Poria Cocos and exerts protective effects against various diseases. PAA treatments also provoked ferroptosis in T-ALL cells with reduced glutathione (GSH) levels and elevated malonaldehyde (MDA) content through inducing autophagic cell death and ferroptosis [55]. Yang et al. provided the first direct evidence that circ_0000745 promoted glycolytic metabolism and cell cycle progression but suppressed the occurrence of ferroptosis and apoptosis of acute lymphoblastic leukemia (ALL) cells *via* orchestrating the miR-494-3p/NET1 axis. That is, the Circ_0000745/miR-494-3p/NET1 axis might serve as a novel potential target for the treatment and diagnosis of ALL as well [56]. Another study found that FBXW7 was adequate to participate in degrading VDAC3 via modulating ubiquitination of cells to promote Erastin-induced ferroptosis during ALL, which could explain the potentially regulatory link between ferroptosis and autophagy. Moreover, Zhu et al. [49] also demonstrated the value and impact of the combination of Erastin and Rapa for ALL management both *in vivo* and *in vitro*.

2.3 Ferroptosis in chronic leukemia

Chronic leukemias are composed of a broad spectrum of subtypes such as including chronic monocytic leukemia, chronic mylocytic leukemia (juvenile, adult, and familial), chronic myelomonocytic leukemia, and chronic lymphocytic leukemia (CLL), which collectively account for lower than 5% of the childhood hematologic malignancies [57–59]. Recently, some studies have revealed the prognostic value of ferroptosis-related genes in chronic leukemia. For instance, Gong et al. indicated that ferroptosis-related genes can be used to stratify CLL patients based on overall survival (OS) (**Table 1**). Meanwhile, they developed a risk signature containing eight ferroptosis-related genes for predicting the OS of CLL patients [60].

Several reports have shown the potential of triggering ferroptosis for chronic leukemia therapy, particularly for eradicating aggressive malignancies that are resistant to traditional therapies [60–63]. For decades, cysteine metabolism has been identified to have a critical role in cancer cell proliferation and survival, and cysteine depletion has been indicated to inhibit cancer growth and induce tumor cell ferroptosis. Furthermore, Liu et al. [64] have recently showed that cysteine depletion can induce ferroptosis in CML cells and TXNRD1 may be a key regulator gene. This illustrates that cysteine metabolism-induced ferroptosis may be a new idea for the treatment of CML except chemotherapy. Meanwhile, Song et al. [65] found that ferroptosis was involved in imatinib mesylate (IMA)-induced cardiotoxicity during the treatment of CML. In detail, they verified that IMA could downregulate Nrf2 expression but upregulate the P53 and TfR expression and thus increase the cellular ROS and iron, which collectively provided evidence for ferroptosis participation in IMA-induced cardiotoxicity and highlighted ferroptosis as a novel target in IMA-exposed patients.

3. Conclusions and perspectives

Ferroptosis is a newly discovered form of regulated cell death. Iron-dependent lipid peroxidation is a major driver of ferroptosis, and ferroptosis may also occur in leukemia. Ferroptosis is critically involved in the pathogenesis of various leukemia, including acute myeloid leukemia, acute lymphoblastic leukemia, chronic myeloid leukemia and chronic lymphocytic leukemia. With ongoing research, prognostic value of ferroptosis-related genes and potential therapeutic strategy for overcoming chemotherapy resistance are likely to become effective therapeutic strategies for leukemia.

Acknowledgements

The authors would like to thank the members of The First Affiliated Hospital of Soochow University, Gansu Provincial Hospital, Chinese Academy of Sciences (CAS), Hefei Institute of Physical Science, and Institute of Health-Biotech, Health-Biotech (Tianjin) Stem Cell Research Institute Co., Ltd., for their technical support. This study was supported by grants from the National Natural Science Foundation of China (82260031, 81700119, and 81900175), Natural Science Foundation of Jiangsu Province (BE2018652, BK20201168, and BK20190181), the project Youth Fund funded by Shandong Provincial Natural Science Foundation (ZR2020QC097), science and technology projects of Guizhou Province (QKH-J-ZK[2021]-107), the project Youth Fund funded by Jiangxi Provincial Natural Science Foundation (S2021QNJJL0277), Jiangxi Provincial Key New Product Incubation Program from Technical Innovation Guidance Program of Shangrao city (2020G002 and 2020K003), Jiangxi Provincial Novel Research & Development Institutions of Shangrao City (2022AB003, 2020AB002, and 2021F013), Natural Science Foundation of Jiangxi Province (20212BAB216073), the 2021 Central-Guided Local Science and Technology Development Fund (ZYYDDFFZZJ-1), and the Non-profit Central Research Institute Fund of Chinese Academy of Medical Sciences (2019PT320005).

Conflict of interest

The authors declare no conflict of interest.

Notes/thanks/other declarations

Not applicable.

Abbreviations

ROS	reactive oxygen species
AML	acute myeloid leukemia
ALL	acute lymphoblastic leukemia
Gpx4	glutathione peroxidase 4
DFP	deferiprone
DFO	deferoxamine
NAC	*N*-acetylcysteine
Fer-1	ferrostatin-1
Lip-1	liproxstatin-1
CML	chronic myeloid leukemia
CLL	chronic lymphoblastic leukemia
OS	overall survival
CR	complete remission
AMPK	AMP-activated protein kinase
HMGB1	high mobility group box 1
HD	Hydnocarpin D
PAA	poricoic acid A
GSH	glutathione
NDGA	nordihydroguaiaretic acid
HSPCs	hematopoietic stem and progenitor cells
IMA	imatinib mesylate
TYP	typhaneoside

Author details

Baoquan Song[1,2,3] and Leisheng Zhang[4,5,6,7,8]*

1 National Clinical Research Center for Hematologic Diseases, Jiangsu Institute of Hematology, The First Affiliated Hospital of Soochow University, Suzhou, China

2 Institute of Blood and Marrow Transplantation, Soochow University, Suzhou, China

3 Key Laboratory of Thrombosis and Hemostasis of Ministry of Health, Soochow University, Suzhou, China

4 Key Laboratory of Molecular Diagnostics and Precision Medicine for Surgical Oncology in Gansu Province and NHC Key Laboratory of Diagnosis and Therapy of Gastrointestinal Tumor, Gansu Provincial Hospital, Lanzhou, China

5 Jiangxi Research Center of Stem Cell Engineering, Jiangxi Health-Biotech Stem Cell Technology Co., Ltd., Shangrao, China

6 Center for Cellular Therapies, The First Affiliated Hospital and Shandong Provincial Qianfoshan Hospital, Shandong First Medical University, China

7 Key Laboratory of Radiation Technology and Biophysics, Hefei Institute of Physical Science, Chinese Academy of Sciences, Hefei, China

8 Institute of Health-Biotech Group, Tianjin, China

*Address all correspondence to: leisheng_zhang@163.com

IntechOpen

© 2022 The Author(s). Licensee IntechOpen. This chapter is distributed under the terms of the Creative Commons Attribution License (http://creativecommons.org/licenses/by/3.0), which permits unrestricted use, distribution, and reproduction in any medium, provided the original work is properly cited.

References

[1] Dixon SJ, Lemberg KM, Lamprecht MR, Skouta R, Zaitsev EM, Gleason CE, et al. Ferroptosis: an iron-dependent form of nonapoptotic cell death. Cell. 2012;**149**(5):1060-1072

[2] Chen X, Kang R, Kroemer G, Tang D. Ferroptosis in infection, inflammation, and immunity. Journal of Experimental Medicine. 2021;**218**(6)

[3] Yu H, Guo P, Xie X, Wang Y, Chen G. Ferroptosis, a new form of cell death, and its relationships with tumourous diseases. Journal of Cellular and Molecular Medicine. 2017;**21**(4):648-657

[4] Su Y, Zhao B, Zhou L, Zhang Z, Shen Y, Lv H, et al. Ferroptosis, a novel pharmacological mechanism of anti-cancer drugs. Cancer Letters. 2020;**483**:127-136

[5] Lee H, Zandkarimi F, Zhang Y, Meena JK, Kim J, Zhuang L, et al. Energy-stress-mediated AMPK activation inhibits ferroptosis. Nature Cell Biology. 2020;**22**(2):225-234

[6] Lei P, Bai T, Sun Y. Mechanisms of ferroptosis and relations with regulated cell death: a review. Frontiers in Physiology. 2019;**10**:139

[7] Jiang X, Stockwell BR, Conrad M. Ferroptosis: mechanisms, biology and role in disease. Nature Reviews. Molecular Cell Biology. 2021;**22**(4):266-282

[8] Chen S, Chen Y, Zhang Y, Kuang X, Liu Y, Guo M, et al. Iron Metabolism and Ferroptosis in Epilepsy. Frontiers in Neuroscience. 2020;**14**:601193

[9] Ursini F, Maiorino M. Lipid peroxidation and ferroptosis: the role of GSH and GPx4. Free Radical Biology & Medicine. 2020;**152**:175-185

[10] Chen X, Li J, Kang R, Klionsky DJ, Tang D. Ferroptosis: machinery and regulation. Autophagy. 2021;**17**(9): 2054-2081

[11] Sun L, Dong H, Zhang W, Wang N, Ni N, Bai X, et al. Lipid peroxidation, GSH depletion, and SLC7A11 inhibition are common causes of EMT and ferroptosis in A549 cells, but different in specific mechanisms. DNA and Cell Biology. 2021;**40**(2):172-183

[12] Perez MA, Magtanong L, Dixon SJ, Watts JL. Dietary lipids induce ferroptosis in caenorhabditiselegans and human cancer cells. Developmental Cell. 2020;**54**(4):447-454 e444

[13] Gao J, Luo T, Wang J. Gene interfered-ferroptosis therapy for cancers. Nature Communications. 2021;**12**(1):5311

[14] Song X, Long D. Nrf2 and ferroptosis: A new research direction for neurodegenerative diseases. Frontiers in Neuroscience. 2020;**14**:267

[15] Dong H, Qiang Z, Chai D, Peng J, Xia Y, Hu R, et al. Nrf2 inhibits ferroptosis and protects against acute lung injury due to intestinal ischemia reperfusion via regulating SLC7A11 and HO-1. Aging (Albany NY). 2020;**12**(13):12943-12959

[16] Zhang Y, Swanda RV, Nie L, Liu X, Wang C, Lee H, et al. mTORC1 couples cyst(e)ine availability with GPX4 protein synthesis and ferroptosis regulation. Nature Communications. 2021;**12**(1):1589

[17] Friedmann Angeli JP, Krysko DV, Conrad M. Ferroptosis at the crossroads of cancer-acquired drug resistance and immune evasion. Nature Reviews. Cancer. 2019;**19**(7):405-414

[18] Proneth B, Conrad M. Ferroptosis and necroinflammation, a yet poorly explored link. Cell Death and Differentiation. 2019;**26**(1):14-24

[19] Qin Y, Qiao Y, Wang D, Tang C, Yan G. Ferritinophagy and ferroptosis in cardiovascular disease: mechanisms and potential applications. Biomedicine & Pharmacotherapy. 2021;**141**:111872

[20] Wang F, Lv H, Zhao B, Zhou L, Wang S, Luo J, et al. Iron and leukemia: new insights for future treatments. Journal of Experimental & Clinical Cancer Research. 2019;**38**(1):406

[21] Newell LF, Cook RJ. Advances in acute myeloid leukemia. BMJ. 2021;**375**:n2026

[22] Estey E, Hasserjian RP, Dohner H. Distinguishing AML from MDS: A fixed blast percentage may no longer be optimal. Blood. 2022;**139**(3):323-332

[23] Liu H. Emerging agents and regimens for AML. Journal of Hematology & Oncology. 2021;**14**(1):49

[24] Zhu HY, Huang ZX, Chen GQ, Sheng F, Zheng YS. Typhaneoside prevents acute myeloid leukemia (AML) through suppressing proliferation and inducing ferroptosis associated with autophagy. Biochemical and Biophysical Research Communications. 2019;**516**(4):1265-1271

[25] Zhang L, Chi Y, Wei Y, Zhang W, Wang F, Zhang L, et al. Bone marrow-derived mesenchymal stem/stromal cells in patients with acute myeloid leukemia reveal transcriptome alterations and deficiency in cellular vitality. Stem Cell Research & Therapy. 2021;**12**(1):365

[26] Koenig KL, Sahasrabudhe KD, Sigmund AM, Bhatnagar B. AML with myelodysplasia-related changes: development, challenges, and treatment advances. Genes (Basel). 24 July 2020;**11**(8):845

[27] Christopher MJ, Petti AA, Rettig MP, Miller CA, Chendamarai E, Duncavage EJ, et al. Immune escape of relapsed AML cells after allogeneic transplantation. The New England Journal of Medicine. 2018;**379**(24):2330-2341

[28] Kloess S, Oberschmidt O, Dahlke J, Vu XK, Neudoerfl C, Kloos A, et al. Preclinical assessment of suitable natural killer cell sources for chimeric antigen receptor natural killer-based "Off-the-Shelf" acute myeloid leukemia immunotherapies. Human Gene Therapy. 2019;**30**(4):381-401

[29] Ye F, Chai W, Xie M, Yang M, Yu Y, Cao L, et al. HMGB1 regulates erastin-induced ferroptosis via RAS-JNK/p38 signaling in HL-60/NRAS(Q61L) cells. American Journal of Cancer Research. 2019;**9**(4):730-739

[30] Wei J, Xie Q, Liu X, Wan C, Wu W, Fang K, et al. Identification the prognostic value of glutathione peroxidases expression levels in acute myeloid leukemia. Ann Transl Med. 2020;**8**(11):678

[31] Zhou F, Chen B. Prognostic significance of ferroptosis-related genes and their methylation in AML. Hematology. 2021;**26**(1):919-930

[32] Huang X, Zhou D, Ye X, Jin J. A novel ferroptosis-related gene signature can predict prognosis and influence immune microenvironment in acute myeloid leukemia. Bosnian Journal of Basic Medical Sciences. 2022;**22**(4):608-628

[33] Song Y, Tian S, Zhang P, Zhang N, Shen Y, Deng J. Construction and validation of a novel ferroptosis-related prognostic model for acute myeloid leukemia. Frontiers in Genetics. 2021;**12**:708699

[34] Wang J, Zhuo Z, Wang Y, Yang S, Chen J, Wang Y, et al. Identification and validation of a prognostic risk-scoring model based on ferroptosis-associated cluster in acute myeloid leukemia. Frontiers in Cell and Development Biology. 2021;**9**:800267

[35] Zheng Z, Wu W, Lin Z, Liu S, Chen Q, Jiang X, et al. Identification of seven novel ferroptosis-related long non-coding RNA signatures as a diagnostic biomarker for acute myeloid leukemia. BMC Medical Genomics. 2021;**14**(1):236

[36] Han C, Zheng J, Li F, Guo W, Cai C. Novel prognostic signature for acute myeloid leukemia: bioinformatics analysis of combined CNV-driven and ferroptosis-related genes. Frontiers in Genetics. 2022;**13**:849437

[37] Zhang X, Zhang X, Liu K, Li W, Wang J, Liu P, et al. HIVEP3 cooperates with ferroptosis gene signatures to confer adverse prognosis in acute myeloid leukemia. Cancer Medicine. 10 May 2022. DOI: 10.1002/cam4.4806. Epub ahead of print. PMID: 35535739

[38] Wei J, Nai GY, Dai Y, Huang XJ, Xiong MY, Yao XY, et al. Dipetidyl peptidase-4 and transferrin receptor serve as prognostic biomarkers for acute myeloid leukemia. Ann Transl Med. 2021;**9**(17):1381

[39] Yu Y, Xie Y, Cao L, Yang L, Yang M, Lotze MT, et al. The ferroptosis inducer erastin enhances sensitivity of acute myeloid leukemia cells to chemotherapeutic agents. Molecular & Cellular Oncology. 2015;**2**(4):e1054549

[40] Du J, Wang T, Li Y, Zhou Y, Wang X, Yu X, et al. DHA inhibits proliferation and induces ferroptosis of leukemia cells through autophagy dependent degradation of ferritin. Free Radical Biology & Medicine. 2019;**131**:356-369

[41] Birsen R, Larrue C, Decroocq J, Johnson N, Guiraud N, Gotanegre M, et al. APR-246 induces early cell death by ferroptosis in acute myeloid leukemia. Haematologica. 2022;**107**(2):403-416

[42] Dong LH, Huang JJ, Zu P, Liu J, Gao X, Du JW, et al. CircKDM4C upregulates P53 by sponging hsa-let-7b-5p to induce ferroptosis in acute myeloid leukemia. Environmental Toxicology. 2021;**36**(7):1288-1302

[43] Yusuf RZ, Saez B, Sharda A, van Gastel N, Yu VWC, Baryawno N, et al. Aldehyde dehydrogenase 3a2 protects AML cells from oxidative death and the synthetic lethality of ferroptosis inducers. Blood. 2020;**136**(11):1303-1316

[44] Lai X, Sun Y, Zhang X, Wang D, Wang J, Wang H, et al. Honokiol induces ferroptosis by upregulating HMOX1 in acute myeloid leukemia cells. Frontiers in Pharmacology. 2022;**13**:897791

[45] Aldoss I, Forman SJ, Pullarkat V. Acute lymphoblastic leukemia in the older adult. Journal of Oncology Practice/ American Society of Clinical Oncology. 2019;**15**(2):67-75

[46] Gregory S. Adult acute lymphoblastic leukemia: treatment and management updates. Seminars in Oncology Nursing. 2019;**35**(6):150951

[47] Jabbour E, Kantarjian H. A new era in the treatment of acute lymphoblastic leukemia. Blood. 2021;**137**(12):1563-1564

[48] Jin L, Tong L. PAQR3 inhibits proliferation and aggravates ferroptosis in acute lymphoblastic leukemia through modulation Nrf2 stability. Immunity, Inflammation and Disease. 2021;**9**(3):827-839

[49] Zhu T, Liu B, Wu D, Xu G, Fan Y. Autophagy regulates VDAC3

ubiquitination by FBXW7 to promote erastin-induced ferroptosis in acute lymphoblastic leukemia. Frontiers in Cell and Development Biology. 2021;**9**:740884

[50] Hong Y, Zhang L, Tian X, Xiang X, Yu Y, Zeng Z, et al. Identification of immune subtypes of Ph-neg B-ALL with ferroptosis related genes and the potential implementation of Sorafenib. BMC Cancer. 2021;**21**(1):1331

[51] Probst L, Dachert J, Schenk B, Fulda S. Lipoxygenase inhibitors protect acute lymphoblastic leukemia cells from ferroptotic cell death. Biochemical Pharmacology. 2017;**140**:41-52

[52] Ishikawa C, Senba M, Mori N. Evaluation of artesunate for the treatment of adult T-cell leukemia/lymphoma. European Journal of Pharmacology. 2020;**872**:172953

[53] Greco G, Schnekenburger M, Catanzaro E, Turrini E, Ferrini F, Sestili P, et al. Discovery of sulforaphane as an inducer of ferroptosis in u-937 leukemia cells: expanding its anticancer potential. Cancers (Basel). 2021;**14**(1)

[54] Lou S, Hong H, Maihesuti L, Gao H, Zhu Z, Xu L, et al. Inhibitory effect of hydnocarpin D on T-cell acute lymphoblastic leukemia via induction of autophagy-dependent ferroptosis. Experimental Biology and Medicine (Maywood, N.J.). 2021;**246**(13):1541-1553

[55] Chen L, Fang W, Liu J, Qi X, Zhao L, Wang Y, et al. Poricoic acid A (PAA) inhibits T-cell acute lymphoblastic leukemia through inducing autophagic cell death and ferroptosis. Biochemical and Biophysical Research Communications. 2022;**608**:108-115

[56] Yang X, Li Y, Zhang Y, Liu J. Circ_0000745 promotes acute lymphoblastic leukemia progression through mediating miR-494-3p/NET1 axis. Hematology. 2022;**27**(1):11-22

[57] Morrison VA. Chronic leukemias. CA: A Cancer Journal for Clinicians. 1994;**44**(6):353-377

[58] Altman AJ. Chronic leukemias of childhood. Pediatric Clinics of North America. 1988;**35**(4):765-787

[59] Young JA. Epidemiology and management of infectious complications of contemporary management of chronic leukemias. Infectious Disorders Drug Targets. 2011;**11**(1):3-10

[60] Gong H, Li H, Yang Q, Zhang G, Liu H, Ma Z, et al. A ferroptosis molecular subtype-related signature for predicting prognosis and response to chemotherapy in patients with chronic lymphocytic leukemia. BioMed Research International. 2022;**2022**:5646275

[61] Zhu S, Liu J, Kang R, Yang M, Tang D. Targeting NF-kappaB-dependent alkaliptosis for the treatment of venetoclax-resistant acute myeloid leukemia cells. Biochemical and Biophysical Research Communications. 2021;**562**:55-61

[62] Pramil E, Herbi Bastian L, Denefle T, Nemati F, Xiao M, Larde E, et al. Targeting chronic lymphocytic leukemia with N-methylated thrombospondin-1-derived peptides overcomes drug resistance. Blood Advances. 2019;**3**(20):2920-2933

[63] Li R, Yin B, Zeng D, Liu Z. A novobiocin derivative, XN4, triggers ferroptosis in gastric cancer cells via the activation of NOX4. Pharmaceutical Biology. 2022;**60**(1):1449-1457

[64] Liu S, Wu W, Chen Q, Zheng Z, Jiang X, Xue Y, et al. TXNRD1: a key regulator involved in the ferroptosis of

CML cells induced by cysteine depletion in vitro. Oxidative Medicine and Cellular Longevity. 2021;**2021**:7674565

[65] Song C, Li D, Zhang J, Zhao X. Role of ferroptosis in promoting cardiotoxicity induced by Imatinib Mesylate via down-regulating Nrf2 pathways in vitro and in vivo. Toxicology and Applied Pharmacology. 2022;**435**:115852

Chapter 4

The "Irony" of Ferroptosis: A Review on Neurological Challenges

Chayan Munshi and Shelley Bhattacharya

Abstract

Ferroptosis in recent days has gained high impact due to its implication in inducing several neurological challenges. Impairment of iron homeostasis (mainly surplus iron deposition) is the key reason for the induction of the ferroptotic cell death. This type of programmed cell death in the neurons can trigger neuropathological abnormalities. Ferroptosis has been given clinical importance, where biomedical researchers are working on the pathological detection of ferroptosis and finding clinical ways to arrest it. In this review, we have elucidated the impact of ferroptotic cell death on the pathophysiology of several neurological challenges.

Keywords: Ferroptosis, cell death, neurological challenges, oxidative stress, iron

1. Introduction

Ferroptosis is recently discovered non-apoptotic, iron-dependent programmed cell death instigated by the upsurge of intracellular lipid reactive oxygen species. Despite of the importance of iron in the body, the proper cellular homeostasis is important. The cellular mechanistic pathways which are related to ferroptosis are majorly iron metabolism and lipid peroxidation. Ferroptosis has been reported to have direct execution for several neurological disorders. As we know that ferroptosis is rather a very new area of research, intricate research is needed to work on the molecular crosstalk between ferroptosis, apoptosis, autophagy, and necrosis. A comparative molecular mechanism of ferroptotic pathophysiology in several neurological diseases needs to be revealed. Furthermore, research on the establishment of ferroptosis as a therapeutic approach for several neurological diseases is also important [1].

The ideal purpose of iron homeostasis in body must be established in the system, which not only transfers iron in the brain but also reduces its concentration, if exceed the optimum level. Irregularities in iron homeostasis, especially for surplus iron, relates to several critical cellular dysfunction and signifies a serious stage for neurodegenerative physiology. Ferroptosis is characterised by iron-dependent oxidative damage in the lipid bilayer. It is mainly instigated by the disparity in the oxidation and anti-oxidation proportion in the cell. Disturbances in iron and lipid metabolism cause excessive accumulation of lipid peroxides within the lipid bilayer, causing oxidative destruction of the cell membrane. Disorder in the cellular antioxidant procedures results in the incapability to remove the lipid peroxides which is generated from the induction of oxidative stress. Eventually which is the reason for the massive annihilation of the lipid bilayer membrane and eventually causes programmed cell death. As accumulation of iron and lipid peroxides play the major roles in the triggering of ferroptosis, thus it can be decreased by the administration of iron chelators or by lipophilic antioxidants in the system. Recent findings have specified its major role in the brain maturation and adverse effects on the nervous system and its pathophysiology eventually initiating neural dysfunctions.

Lipid hydroperoxides are formed in the process of lipid peroxidation [2]. Glutathione peroxidase (GPX4) is a glutathione (GSH) dependent enzyme that acts to reduce lipid hydroperoxides (L-OOH) to from lipid alcohols (L-OH). This can inhibit the iron induced formation of toxic lipid reactive oxygen species (L-ROS). GSH is a cofactor of GPX4 and effectively sustains the GPX4 level through cystine/glutamate antiporter system known as the Xc- system. The inhibition of Xc- system or GSH synthesis, or GPX4 activity will eventually initiate the accumulation of lipid peroxides and initiation of ferroptotic cell death. The dysfunction of iron metabolism system, iron uptake (transferrin receptor), iron export (ferroportin), iron accumulation (ferritin) induces surplus of iron load and which results in the catalysis of hazardous L-ROS production [3]. The antiporter is a 12-pass transmembrane subunit (SLC7A11) where anionic cystine enters inside the cell via facilitated diffusion and in exchange, anionic glutamate moves out of the cell by facilitated diffusion. This antiporter also has 1-pass transmembrane subunit (SLC3A2), connected to the transporter by a disulfide bond. The capability of glutamate analogues to provoke ferroptosis in neurons is straightway correlated with their capability to inhibit cystine uptake in the cell. Cysteine in the extracellular domain is quickly oxidised to cystine where two molecules of cysteines are linked covalently by a disulfide bond. After transportation inside the cell, cystine is reduced to cysteine by glutathione reductase. Inhibition of cystine uptake results in the exhaustion of cysteine and eventually related exhaustion of GSH [4].

In this chapter, we have reviewed the impact of ferroptotic cell death on the pathophysiology of several neurological challenges.

2. Mechanistic overview of the pathophysiological manifestation during ferroptosis-induced neurological challenges

The propagation of ageing in nervous system, induces clinical conditions like Alzheimer's disease (AD), Parkinson's disease (PD), Huntington's disease (HD), anxiety, depression, stroke, and traumatic brain injury which appears to be predominant in a population. The molecular mechanism of neurological disorders is multifarious, and there is lack of applicable therapeutic protocols. Molecules, manifesting ferroptosis-induced neurological disorders, such as reactive oxygen species (ROS), nuclear factor erythroid 2-related factor 2/antioxidant response element (Nrf2/ARE), iron ion (Fe^{2+}), nicotinamide adenine dinucleotide phosphate (NADPH), NADPH oxidase (NOX), play crucial role in the pathophysiology. In the cells, due to the activity of acyl-CoA synthetase long chain family member 4 (ACSL4), cytochrome p450 oxidoreductase (POR), lipoxygenases (ALOX), polyunsaturated fatty acids (PUFA) produce phospholipid hydroperoxides (PLOOH) over a sequence of biochemical reactions, eventually from which, PLOO is generated.

This results in lipid peroxidation. GPX4 activity can reduce PLOOH to PLOH to constrain lipid peroxidation occurrence. Mitochondria generates reactive ROS through, which causes oxidative stress. Fe^{2+} is oxidised to ferric ion (Fe^{3+}) after absorption in the duodenum. Fe^{3+} enters the cell by combining with transferrin (Tf) and transferrin receptor (Tfr) to procedure a complex. Iron decomposition from endosome can induce ferroportin (FPN) protein activation on the cell membrane, where iron ions enter iron pool. Tf-Tfr complex triggers for the next cell cycle. During neurological disorders, excessive iron produces a large volume of ROS, which cause ferroptosis [5].

Neurodegeneration, cardiovascular disease, and diabetes, where ferroptosis might provoke the neuronal, cardiomyocyte, and β-cell loss. Regarding ferroptosis research, the transcription factor Nrf2 and its transcriptional target genes play in the inhibition process or in certain situations activating of the ferroptotic cascade. This concludes that cautious attention should be given in terms of Nrf2 pathway, which signify practical targets to recruit ferroptosis in tumour cell types, without damaging their normal cell types. Another therapeutic option is using the activation of Nrf2 or other essential anti-ferroptotic moderators in terms of avoiding ferroptosis. In homeostatic circumstances, Nrf2 is ubiquitylated and embattled for proteasomal.

degradation by a KEAP1-CUL3-RBX1 E3 ubiquitin ligase complex. During the pathophysiology of oxidative stress, or mutations occurred in Nrf2 or KEAP1/CUL3, the Nrf2 is not degraded, which in turn allows nuclear translocation and activation of antioxidant response element encompassing genes. Nrf2 is involved in iron metal metabolism, along with the detoxification system of the cell through glutathione synthesis, all of these play a crucial key role in the inhibition of ferroptosis. In the circulation (blood), Fe^{3+} is transported by Tf. Fe^{3+}-Tf is bound by tfr1 and endocytosed. In the endosome, the acidic pH promotes detachment of Tfr and Fe^{3+}. This is further reduced to Fe^{2+} by the activity of metalloreductase STEAP3. Fe^{2+} is further transported to the cytosol by divalent metal transporter 1 (DMT1), donating to the iron pool. FPN1 plays a role in exporting of Fe^{2+} out of the cell by incorporating it into iron-containing proteins or storing by ferritin as Fe^{3+}. Ferritinophagy is the autophagic mechanism which regulates the degradation of ferritin through nuclear receptor coactivator 4 (NCOA4). The degradation of ferritin, results in the reduction of Fe^{3+} to Fe^{2+} by STEAP3 and eventually exported from the lysosome to the cytosol by DMT1 to contribute to the iron pool. Notably, the physiology of iron metabolism, storage, and transport, along with ferritinophagy are transcriptionally controlled by Nrf2 [6].

Both oxidative and nitrosative stress can unfavourably disturbs the mechanistic pathways and effective proteins regulating cellular iron homeostasis, like iron controlling protein/iron response element system, and can eventually be a basis of unusually high levels of iron and a cause of lethal levels of lipid membrane peroxidation. Moreover, neuroinflammation governs the upregulation of bivalent metal transporter-1 on the surface of astrocytes, microglia, and neurones, which make them extremely sensitive to excess iron in the occurrence of elevated levels of non-transferrin bound iron, therefore, initiation of iron mediated neuropathology occurs. Mechanisms regulating the iron homeostasis physiology and the effectiveness of ferritin and mitochondria are important. Negative regulation of ferroptosis by GSH, GPX4, the cystine/glutamate antiporter system, heat shock protein 27 and Nrf2 is crucial. The possible role of deglycase (DJ-1) inactivation in the reduction of ferroptotic cell death is simultaneously critical. Therapeutic approach in terms of coenzyme Q10, iron chelation therapy, deferiprone, deferoxamine (desferrioxamine) and deferasirox, and N-acetylcysteine is of high clinical importance [7].

3. Ferroptotic influence on the manifestation of neurological challenges

Iron overload in cell (dysregulation of iron homeostasis) is contemplated to be a precarious situation for neurodegeneration. The current findings, emphasise β-amyloid, tau proteins, α-synuclein, and demyelination process connected to ferroptosis which induces neurodegeneration. The theory built on the possible role of dysregulation of iron homeostasis and ferroptosis in the pathophysiology of neurodegenerative diseases can be further supported by clinical experiments and epidemiological analyses [8].

Despite of the fact that ferroptosis was first defined in cancer cells, however, developing indications, include this cell death process with cerebral ischemia and brain haemorrhage also. Neonatal brain injury is a significant reason for the developmental damage and everlasting neurological insufficiencies. Different cell death processes, including iron-dependent ferroptotic pathways, have been identified for neonatal brain injury. Iron chelators and erythropoietin have been acknowledged as neuroprotective agents against neonatal brain injury. Generally, ferroptosis is principally defined through activators and inhibitors. Ferroptosis in adults are reported to generate ischemic and intraventricular haemorrhage-induced neuronal cell death. The inhibition of ferroptosis decreases the rate of neuronal death and behavioural abnormalities. Contemplating the recent paradigms in ferroptotic research, investigation on the relation between neonatal brain injury and ferroptosis should be considered seriously [9].

The propagation of ferroptosis incorporates the pathophysiology of autophagy as well as the inclusion of the activities of well-studied proteins such as Nrf2, p53 etc. The manifestation and regulatory molecular pathways of ferroptosis are constantly developing. There are indications that ferroptosis and its correlated genes may be concerned in a sequence of neural maturation, maintenance, and ageing physiology. In turn these genes can play a critical role in the pathophysiology of neurodegenerative diseases, neurological disorders, strokes, epilepsy, brain tumours etc. Investigating the incidence and expansion of ferroptosis in the brain tumours and endeavouring to stimulate tumour cell death with ferroptosis inducers are expected to collaborate with traditional tumour therapeutics and immunotherapy protocols. However, in the perspective of glioma treatment by endorsing ferroptosis, it is possible for the destruction of neuronal cells simultaneously, thus provoking the collective neurodegenerative diseases, stroke, and other neurological disorders, which will result to the similar indications in glioma patients [10].

In ferroptosis, a cloud of pharmacological modulators has been discovered considering the target proteins involved in iron homeostasis, in terms of origination and reduction of lipid peroxides or cystine import and GSH metabolism. Several machineries of the ferroptosis cascade are target genes of the transcription factor Nrf2, representing its analytical role in facilitating the ferroptotic reaction. Ferroptosis, is controlled by various cellular metabolic pathways. Research on the effect of ferroptotic cell death in inducing numerous neurological disorders has gained acceleration nowadays. Genetic regulation behind ferroptosis-induced neurological disorders and the probable functioning of ferroptosis in the development of brain is of serious concern. There are reports on 42 ferroptosis genes, which play crucial roles in the brain development and the gene co-expression system for the human dorsolateral prefrontal cortex development, where cluster of 22 genes actively participate. 12 genes out of these 22 genes are considered for the conservation of elementary cellular functions (non-transitional), which include RNA processing.

The rest 10 genes with postnatal line are effective for the upgradation of patterns in neuron and glial cells. Stress affects the differential gene expression pattern during the process of brain development [11].

Cumulative substantiation designates a probable connection between neuroinflammation and neurological disorders, including AD, PD, HD, and stroke. Ferroptosis can possibly explain this connection. Research have shown that disorders of iron homeostasis, glutamate excitative toxicity, L-ROS, and some other factors related to ferroptosis can be detected in several neurological disorders which is caused by neuroinflammation. Convincing indication regarding the damage-associated molecular pattern molecules, like ROS, generated during the pathophysiological process of ferroptosis, trigger glial cells by stimulating neuroimmune pathways and then generate a sequence of inflammatory factors which initiate neurological disorders. Complicated biochemical reactions occur throughout ferroptosis. Activated microglia, reactive astrocytes, invasive T-cells, and overproduction of inflammatory molecules, establish the neuroinflammatory response. During the early phase, acute inflammatory responses cause trauma in the central nervous system, which can have a defensive role, restraining the strictness of the injury thus augmenting neuronal repair. If the acute inflammatory response does not decrease adequately, it will be directed into chronic inflammation which can be uncontrollable. In this situation, glial cells incline to intensify oxidative stress on neurons. Neuroinflammation is not essential during the early stage of neurological disorders, however, a constant inflammatory response can produce aggravation of the diseases. The detail pathophysiology of ferroptosis and its connection with neuroinflammation, have been understood in a rudimentary level. The practice of inhibitors of ferroptosis in investigational animal models can improve the rigorousness of the neural diseases. However, medications targeting ferroptosis can play a critical role in the medical treatment of chronic neuroinflammatory diseases. This needs rigorous clinical trials [12].

The active equilibrium of cardiomyocytes and neurons is vital to continue the normal physiological functions of heart and brain. If unnecessary cell death occurs in the tissues, severe cardio-cerebrovascular diseases (CCVD) like, hypertension, myocardial infarction, and ischemic stroke happens. The mechanistic regulation of cell death possesses a key role in endorsing these diseases. Worldwide, major mortalities and morbidities occur due to CCVDs. Excess of iron has been established to be a crucial element for pathogenic response in cardiocerebrovascular toxicity and manifest diverse CCVDs, thus ferroptosis, has received serious attention for its pathophysiology for CCVDs. Many studies have shown that ferroptosis occurs in atherosclerosis, heart failure, diabetic cardiomyopathy, hypertensive brain injury, ischemic stroke, and myocardial infarction. Inhibitors of ferroptosis may stop these diseases by inhibiting the ferroptotic pathway both in cardiac tissue and neurons. Cardio-cerebrovascular cell death is a central pathophysiological procedure and in fact, ferroptosis strikes throughout the CCVDs. However, in terms of abridged level of ferroptosis inhibitors (like GSH, GPX4, and Nrf2) and alterations in the gene expression levels, which are known to be expressed during CCVDs, the existing assays are not totally appropriate for predictable and routine clinical diagnosis. Damaged mitochondrial structure is an important feature of ferroptosis. Growing figure of inducers and inhibitors of ferroptosis have been showed. However, the best therapeutic agent among these inducers and inhibitors are still to known properly. Ferroptosis leads to pathophysiological alterations like inflammation and endoplasmic reticulum stress. However, it is a complex issue to explain the whole pathway across which ferroptosis works with the initiation of ischemia and hypoxia, iron discharges through the upregulation of heme oxygenase 1 (Hmox1), thus stimulating ferroptosis, resulting in myocardial pathologic modification and myocardial cell damage. Heart failure enthused by enhancement of the iron pool resulting in the surplus iron and the incidence of ferroptosis, eventually which leads to myocardial edema, arrhythmia, and cardiomyocyte cell death [13].

Intracerebral haemorrhage (ICH) is another critical medical condition with high morbidity and mortality. Brain injury due to ICH is primarily recognised due to oxidative stress and haemoglobin lysate (which include iron), indicates unalterable harm to neurons. Therefore, ferroptosis has become a recent paradigm in neuronal cell death research after ICH [14].

The blood–brain barrier (BBB) is crucial in regulating the homeostasis within the CNS. Brain microvascular endothelial cells are effectively arrayed to make the vessel walls and have tight junction complexes that restrict the paracellular pathways of BBB. These walls effectively controls the movement of ions, molecules, and cells between the blood and the brain. This is extremely important for the protection of the neural tissues in the brain from hazardous toxins and pathogens. Primary damage due to the ill functioning of BBB can damage the tight junctions, transport proteins and leukocyte adhesion molecules, which can cause brain edema, imbalance in ion homeostasis, changed signalling pathways and immune infiltration, leading eventually to neuronal cell death. Several neurological disorders can happen due to BBB dysfunction. Ferroptosis can play a key role in BBB dysfunction [15].

Severe central nervous system (CNS) injuries, like stroke, traumatic brain injury, and spinal cord injury is a serious cause of concern for clinicians due to high morbidity and mortality. In fact, the therapeutic strategies for these diseases are not sufficient some time. Oxidative stress, neuroinflammation, excitotoxicity, and programmed cell death (including ferroptosis) plays crucial roles in the pathophysiology of acute CNS damages. Reports develop relations between acute CNS injuries and ferroptosis. Pharmaceutical agents, such as iron chelators, ferrostatin-1 (Fer-1), and liproxstatin-1 (Lip-1), can have inhibitory effect on the ferroptosis and may have neuroprotective capabilities even after acute CNS injuries. Till date, edaravone is the single approved medicine with accepted clinical effectiveness and safety for the CNS injury [16].

4. Conclusion

Recent research on programmed cell death complemented by ferroptosis, is indicating to find novel theories on ferroptotic mechanism to design therapeutic protocols for ferroptosis-related neurological diseases. A medical perception is indeed necessary for the treatment strategy to combat the dysregulation of iron homeostasis and/or inhibition of ferroptosis to reduce the rate of neurodegenerative pathophysiology induced by ferroptosis. Our knowledge on ferroptosis is still at the base level. It is extremely important to depict a ferroptotic biomarkers for biomedical identification. Ferroptosis has distinctive process which has produced abundant chemotherapeutic potentials for treating cancers. The concrete pathophysiological implication of ferroptotic pathway, encompassing reasonable translational methods is still evolving. Growing data propose that the inhibition of ferroptosis may efficiently avoid neuronal diseases.

Author details

Chayan Munshi[1,2]* and Shelley Bhattacharya[3]

1 International Health Management Programme, Berlin School of Business and Innovation, Berlin, Germany

2 Ethophilia (An Autonomous Research Group), Santiniketan, India

3 Department of Zoology, Visva Bharati University, Santiniketan, India

*Address all correspondence to: chayanbio@gmail.com

IntechOpen

© 2022 The Author(s). Licensee IntechOpen. This chapter is distributed under the terms of the Creative Commons Attribution License (http://creativecommons.org/licenses/by/3.0), which permits unrestricted use, distribution, and reproduction in any medium, provided the original work is properly cited.

References

[1] Lei J, Chen Z, Song S, Sheng C, Song S, Zhu J. Insight into the role of ferroptosis in non-neoplastic neurological diseases. Frontiers in Cellular Neuroscience. 2020;**14**:231

[2] Forcina GC, Dixon SJ. GPX4 at the crossroads of lipid homeostasis and ferroptosis. Proteomics. 2019;**19**(18):1800311

[3] Yao MY, Liu T, Zhang L, Wang MJ, Yang Y, Gao J. Role of ferroptosis in neurological diseases. Neuroscience Letters. 2021;**747**:135614

[4] Ratan RR. The chemical biology of ferroptosis in the central nervous system. Cell Chemical Biology. 2020;**27**(5):479-498

[5] Li J, Jia B, Cheng Y, Song Y, Li Q, Luo C. Targeting molecular mediators of Ferroptosis and oxidative stress for neurological disorders. Oxidative Medicine and Cellular Longevity. 2022;**2022**:1-14

[6] Anandhan A, Dodson M, Schmidlin CJ, Liu P, Zhang DD. Breakdown of an ironclad defense system: The critical role of NRF2 in mediating ferroptosis. Cell Chemical Biology. 2020;**27**(4):436-447

[7] Morris G, Berk M, Carvalho AF, Maes M, Walker AJ, Puri BK. Why should neuroscientists worry about iron? The emerging role of ferroptosis in the pathophysiology of neuroprogressive diseases. Behavioural Brain Research. 2018;**341**:154-175

[8] Viktorinova A, Durfinova M. Mini-review: Is iron-mediated cell death (ferroptosis) an identical factor contributing to the pathogenesis of some neurodegenerative diseases? Neuroscience Letters. 2021;**745**:135627

[9] Wu Y, Song J, Wang Y, Wang X, Culmsee C, Zhu C. The potential role of ferroptosis in neonatal brain injury. Frontiers in Neuroscience. 2019;**13**:115

[10] Zhou Y, Lin W, Rao T, Zheng J, Zhang T, Zhang M, et al. Ferroptosis and its potential role in the nervous system diseases. Journal of Inflammation Research. 2022;**15**:1555

[11] Kim SW, Kim Y, Kim SE, An JY. Ferroptosis-related genes in neurodevelopment and central nervous system. Biology. 2021;**10**(1):35

[12] Cheng Y, Song Y, Chen H, Li Q, Gao Y, Lu G, et al. Ferroptosis mediated by lipid reactive oxygen species: A possible causal link of Neuroinflammation to neurological disorders. Oxidative Medicine and Cellular Longevity. 2021;**2021**

[13] Luo MY, Su JH, Gong SX, Liang N, Huang WQ, Chen W, et al. Ferroptosis: New dawn for overcoming the cardio-cerebrovascular diseases. Frontiers in Cell and Developmental Biology. 2021;**9**:1-18

[14] Bai Q, Liu J, Wang G. Ferroptosis, a regulated neuronal cell death type after intracerebral hemorrhage. Frontiers in Cellular Neuroscience. 2020;**14**:591874

[15] Zhao Y, Liu Y, Xu Y, Li K, Zhou L, Qiao H, et al. The role of Ferroptosis in blood–brain barrier injury. Cellular and Molecular Neurobiology. 2022;**42**:1-14

[16] Shen L, Lin D, Li X, Wu H, Lenahan C, Pan Y, et al. Ferroptosis in acute central nervous system injuries: The future direction? Frontiers in Cell and Developmental Biology. 2020;**8**:594

Chapter 5

The Roles of Iron and Ferroptosis in Human Chronic Diseases

Yanbo Shi, Junyong Zhang, Kaitao Luo, Sunfeng Pan, Hanqiang Shi, Lie Xiong and Shuqin Du

Abstract

Ferroptosis, an iron-dependent novel type of cell death, has been characterized as an excessive accumulation of lipid peroxides and reactive oxygen species. A growing number of studies demonstrate that ferroptosis not only plays an important role in the pathogenesis and progression of chronic diseases, but also functions differently in different diseases. As a double-edged sword, activation of ferroptosis could potently inhibit tumor growth and increase sensitivity to chemotherapy and immunotherapy in various cancer settings. Therefore, the development of more efficacious ferroptosis agonists or inhibitors remains the mainstay of ferroptosis-targeting strategy for cancer therapeutics or cardiovascular and cerebrovascular diseases and neurodegenerative diseases therapeutics.

Keywords: iron metabolism disorder, ferroptosis, tumor, neurodegeneration, vascular diseases

1. Introduction

Chronic diseases are non-infectious, of long duration, and are persistent diseases. These diseases mainly include cardiovascular and cerebrovascular diseases, tumors, diabetes, and chronic respiratory diseases, etc. According to the statistics of WHO, approximately 41 million people worldwide die of chronic diseases, accounting for 73.6% of all deaths in 2021 [1]. The occurrence and development of chronic diseases are not only complicated in pathological mechanisms, but also affected by many factors such as genetics, understanding of the molecular mechanisms underlying the pathogenesis of chronic diseases is helpful for diagnosis and treatment. In 2012, a new way of cell death, ferroptosis, was discovered and reported, to some extent, the discovery is a milestone in the study of cell death; it provides a new perspective to study the occurrence, development, and prevention of chronic diseases.

As an essential trace element, iron is present in nearly all forms of life and involved in various of biological processes, including respiration, oxygen transport, intermediary metabolism, gene regulation, and nucleotide synthesis and repair [2, 3]. However, dysregulated iron homeostasis leads to common hematological, metabolic, and neurodegenerative diseases.

Ferroptosis is an iron-dependent cell death, it is different from apoptosis, pyroptosis, or necrosis in morphology, genetics, metabolism, and molecular biology [4]. The morphology is mainly manifested as mitochondrial swelling, increased membrane density, smaller volume, decreased number of cristae, increased lamellar phenotype, and increased autophagosomes, etc. Molecular biology is mainly manifested as glutathione (glutathione, GSH) depletion or inactivation of glutathione peroxidases (GPX4), increased intracellular free iron content, and increased production of reactive oxygen species (ROS), etc., ultimately manifested as the accumulation of toxic lipid hydroperoxides in cells [4]. Its main characteristics are excessive accumulation of lipid peroxides and reactive oxygen species [5]. Since the production of toxic lipid hydroperoxides mostly depends on ferrous iron, and specific iron chelators can inhibit iron-disturbance-mediated ferroptosis. Therefore, iron metabolism and lipid peroxidation play an important regulating role in ferroptosis pathways [6, 7], the possible molecular mechanism is shown in **Figure 1**.

Ferric ions in the circulation are bound to transferrin and transported into the cells through transferrin receptor 1 (TFR1), which is located on the cell membrane. After being reduced to divalent iron in the cell, ferric iron is transported by divalent metal transport1 (DMT1) and released into the cytoplasmic iron pool, excess iron is stored

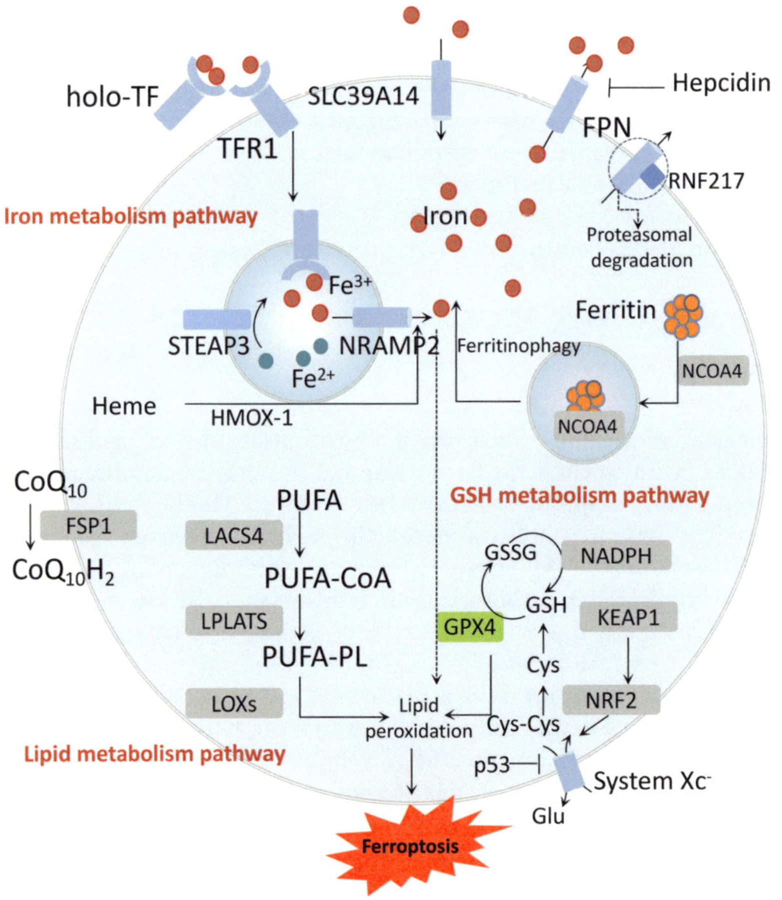

Figure 1.
Molecular mechanism of ferroptosis.

in ferritin. Some studies showed that ferritin selective autophagy promotes ferritin getting into autophagosomes through the nuclear receptor co-activator 4 pathway and results in the releasing of free iron [8].

Generally, it is believed that excess iron causes ferroptosis mainly through reactive oxygen species produced by the Fenton reaction, application of iron chelators can effectively inhibit ferroptosis [10]. Lipid peroxides play the role of agents in the process of ferroptosis, phosphatidyl ethanolamine (PE) is the substrate of choice for lipid oxidation. Therefore, hydrogen peroxide-PE (OOH-PE) is considered to be the main signal of ferroptosis [9]. In the process of lipid peroxidation accumulation, NADPH oxidase, lipoxygenase, Acyl-CoA long-chain family member 4(ACSL4), and lysophosphatidyl cholinyltransferase 3 may play important roles in the occurrence and development of ferroptosis [10–12]. The commonly used ferroptosis inhibitors ferrostatin-1(Fer-1), liproxstatin-1(Lip-1), and small-molecule compounds such as vitamin E inhibit ferroptosis mainly by scavenging lipid peroxides [13].

Glutathione peroxidase 4 (GPX4) is a crucial enzyme in the regulation of ferroptosis; it catalyzes the reduction of lipid peroxides, converts OOH-PE into OH-PE, further suppressing the occurrence of ferroptosis [14]. Small-molecule compounds such as RSL3 and ML162 can inhibit GPX4 activity, lead to the accumulation of fatty acid free radicals, eventually lead to ferroptosis. Reduced glutathione is coenzyme factor of GPX4, the rate-limiting step in its synthesis is the absorption of cystine [15]. Cystine/glutamate transporter consists of the membrane transporters solute carrier family 7 members of 11 (SLC7A11) and regulatory proteins across the membrane of solute carrier family 3 member 2 (SLC3A2); it can transport cystine into the cell and excrete the same amount of glutamate at the same time [16], serves as an important regulator of ferroptosis. Small molecules such as erastin can inhibit glutamic acid/cystine reverse transporter, causing ferroptosis. Wang lab first reported that gene Slc7a11 knockout can promote the occurrence of ferroptosis in mice [17]. Additionally, a novel ferroptosis inhibiting factor 1 (FSP1) was discovered independently in two labs recently, in their studies, NADPH was used to reduce ubiquinone (CoQ_{10}) to ubiquinol ($CoQ_{10}H_2$), leading to the reduction of lipid peroxidation of cell membranes, thereby inhibiting ferroptosis [18, 19]; these findings provide an important basis of developing drugs targeting ferroptosis.

Although the specific mechanisms of ferroptosis are not fully clarified, with the deepening of the research, researchers gradually found that ferroptosis plays an important role in the development of major chronic diseases [20]. So far, a growing evidence showed that ferroptosis is involved in the pathophysiological process of neurodegenerative diseases, tumors, ischemia–reperfusion injury, kidney injury, and other diseases; recent studies have found that ferroptosis plays an important role in cardiovascular disease. Here, we summarized the latest research progress of ferroptosis in cancer, neurodegenerative diseases, and cerebrovascular diseases, to provide new ideas and strategies for the prevention and treatment of major chronic diseases.

2. Ferroptosis and tumor

Tumor cells can proliferate by avoiding cell death, and apoptosis, necrosis, autophagy also played important roles in the development of tumor. In **Table 1**, we summarized the latest research progress of molecular mechanism studies on ferroptosis in common tumors, to provide a series of potential new targets for tumor prevention and control.

Tumor type	Related studies	Reference
Hepatocellular carcinoma	Sorafenib induces ferroptosis in hepatocellular carcinoma cells	[21]
	Retinoblastoma protein-deficient hepatocellular carcinoma cells are more sensitive to sorafenib-induced ferroptosis	[22]
	Activation of p62-Keap1-NRF2 leads to ferroptosis resistance in hepatocellular carcinoma cells	[23–26]
	The expression levels of SLC7A11, Rb, and MT1 are related to the prognosis of patients with hepatocellular carcinoma	[22, 27, 28]
Pancreatic cancer	Artesunate induces ferroptosis in pancreatic ductal adenocarcinoma cells	[29, 30]
	Piperamide induces ferroptosis in pancreatic ductal adenocarcinoma cells	[31]
	Combination of piperamide, Cotylenin A, and sulfasalazine effectively inhibits pancreatic cancer via ferroptosis	[31]
Renal cell carcinoma	Compared with other tumor cells, renal clear cell carcinoma cells are more sensitive to ferroptosis induced by glutathione peroxidase 4 inhibition	[32]
	HIF-2α-HILPDA pathway regulates the sensitivity of renal clear cell carcinoma cells to ferroptosis	[33]
	TAZ/EMP1/NOX4 pathway regulates the sensitivity of renal clear cell carcinoma cells to ferroptosis	[34, 35]
Breast cancer	Siramesine combined with apatinib upregulates iron levels and induces ferroptosis in breast cancer cells	[36]
	Mucin 1C subunit, SLC7A11, and CD44v form a complex to upregulate reduced glutathione expression and make triple-negative breast cancer cells resistant to ferroptosis	[37]
	Sulfasalazine inhibits the growth of glutamine auxotrophic triple-negative breast cancer cells	[38]
	SLC7A11 is closely related to drug resistance and metastasis of triple-negative breast cancer cells	[39]
	Transferrin receptor expression level is associated with breast cancer prognosis	[40, 41]
Bladder tumor	Ferritin phagocytosis releases intracellular free iron and induces ferroptosis and inhibits bladder tumors.	[44]

SLC7A11: solute carrier family 7 member 11; Rb: retinoblastoma; MT1: metallothionein 1.

Table 1.
Ferroptosis in cancer.

2.1 Ferroptosis and hepatocellular carcinoma (HCC)

Targeting ferroptosis is a potential mechanism in the treatment of hepatocellular carcinoma (HCC) [21]. Retinoblastoma (Rb)-deficient cancer cells are more sensitive to ferroptosis induced by sorafenib [22], it is likely that it enhances the oxidative stress response by affecting the concentration of reactive oxygen species in mitochondria. Furthermore, Sun et al. found that nuclear factor E2–related factor 2 (NRF2) protects hepatocellular cancer cells from ferroptosis induced with sorafenib, indicating that targeting p62-Keapl-NRF2 pathway may overcome sorafenib resistance in hepatocellular carcinoma cells [23–26]. Additionally, NRF2 also induces expression of metallothionein1G (MT-1G), an important negative regulator of ferroptosis, through the cythionase pathway, leading to sorafenib resistance in cancer cells [27]. CDGSH [Fe-S]-containing domain 1 can protect the mitochondria from ferroptosis in hepatocellular cancer cells and can be upregulated by erastin in an iron-dependent manner [28]. Moreover, $p53^{S47}$ mutation carried by hepatocellular carcinomas causes ferroptosis tolerance by inhibiting ASCL4 [28].

In hepatocellular carcinoma, the expression of ferroptosis-related genes is related to the prognosis of patients, the mRNA expression levels of *SLC7A11* in HCC tissues and adjacent normal tissues were compared between 130 cases of hepatocellular carcinoma (HCC) tissues and adjacent normal tissues by Kinoshita et al. [29], it showed that the expression of *SLC7A11* in HCC tissues was significantly higher than that in normal tissues, the survival time and disease-free survival time of liver cancer patients were significantly shorter than those of *SLC7A11* low expression of hepatocellular carcinoma patients. In addition, in the process of sorafenib treatment of hepatocellular carcinoma patients, the high expression of Rb and MT1 is also associated with the poor prognosis of patients [22, 27].

2.2 Ferroptosis and pancreatic cancer

The main mechanism of pancreatic cancer is that mutated KRAS gene reprograms pancreatic ductal adenocarcinoma (PDAC) cells to a state that is highly resistant to apoptosis. Artesunate can induce tumor cell apoptosis by generating reactive oxygen species [30]. Eling et al. found that artesunate can induce ferroptosis in PDAC cells with KRAS mutations and the process can be effectively inhibited by Fer-1 [31]. Yamaguchi et al. found that the natural product piperamide can induce ferroptosis in tumor cells by promoting the generation of reactive oxygen species, and its antitumor effect can be inhibited by antioxidants, ferroptosis inhibitors, and iron chelators [32]. The combined use of piperamide, Cotylenin A (a plant growth regulator), and sulfasalazine has a good synergistic effect on pancreatic cancer. These results suggest that ferroptosis inducers are expected to be used in the treatment of pancreatic cancer.

2.3 Ferroptosis and renal cell carcinoma

Renal cell carcinoma originates from the renal parenchyma urothelial system and is a highly malignant tumor in the urinary system. Yang et al. [33] found that GPX4 is a key regulator of the ferroptosis signaling pathway in clear cell renal cell carcinoma. When compared with the other tumor cell from other tissues (lung cancer, colon cancer, central nervous system, melanoma, ovarian cancer, breast cancer, and leukemia), renal cells were more sensitive to ferroptosis induced with inhibition of GPX4. Renal cancer cells are induced by the hepatocyte factor -1β-1-acylglycerol-3 phosphate oxyacyltransferase 3 (AGPAT3) axis and the HIF-2α-HILPDA pathway, which can induce polyunsaturated fatty acyl lipid–enriched cells state, thereby increasing its susceptibility to ferroptosis [34]. Recently, Yang et al. [35] found that the sensitivity of renal cancer cells to ferroptosis is regulated by cell density and transcriptional regulator 1 (TAZ)-TAZ regulation of epidermal membrane protein 1 (EMP1)/NOX4 pathway [34, 35], suggesting that TAZ is a potential therapeutic target for ferroptosis.

2.4 Ferroptosis and breast cancer

Breast cancer is derived from breast epithelial tissue. Ma et al. [36] found that the lysosomal interfering agent siramesine and the tyrosine kinase inhibitor lapatinib can disrupt the iron homeostasis in breast cancer cells to generate reactive oxygen species and induce cell ferroptosis, and overexpression of trasnferrin receptor 1 (TfR1) or iron chelators can reduce siramesine and lapatinib-induced reactive oxygen species. Some studies have shown that the formation of a complex between mucin 1C subunit and SLC7A11 can upregulate the expression of reduced glutathione and inhibit ferroptosis in triple-negative breast cancer cells [37]. Timmerman et al. [38] found a subpopulation of glutamine auxotrophic triple-negative breast cancer cells that were highly dependent on SLC7A11 acquires cystine for glutamine metabolism. The SLC7A11 inhibitor sulfasalazine can inhibit tumor growth by promoting ferroptosis. In addition, the activities of SLC7A11 and glutamate/cystine antiporter can be regulated by the Keap1/NRF2 redox pathway. Lanzardo et al. [39] believed that SLC7A11 is closely related to drug resistance and metastasis of triple-negative breast cancer cells. The increased expression of TFR1 in breast cancer cells is negatively correlated with the expression of estrogen receptor, and the high expression of TFR1 in breast cancer tissue is associated with poor prognosis of patients [40, 41].

2.5 Ferroptosis and bladder tumors

Intracellular iron concentration is closely related to the progression of bladder tumors. Martin-Sanchez et al. [42] analyzed the relationship between intracellular iron concentration and bladder tumor proliferation and found that when transferrin combined with iron, the free iron decreased in tumor cells, which was conducive to the proliferation of bladder cancer cells. When the application of gallium (Ga) to transferrin interferes with the binding of iron to transferrin, the intracellular free iron increases and thus inhibits the proliferation of bladder tumor cells. Mazdak et al. [43] found that the serum iron level of patients with bladder tumors was significantly lower than that of the normal control group, suggesting that the decreased serum iron level may be an important reason for the occurrence of bladder tumors. Tang et al. [44] proposed the phenomenon of ferritin phagocytosis, which releases intracellular free iron through ferritin and increases the content of intracellular free iron, which may play a role in inhibiting bladder tumors. The results suggest that activating ferroptosis can achieve ideal therapeutic effect on bladder tumors. These studies suggest that increasing intracellular iron concentration may help to inhibit bladder tumor progression.

2.6 Tumor-associated ferroptosis regulatory protein

2.6.1 SLC7A11 regulates ferroptosis

SLC7A11(also known as xCT) is the substrate-specific subunit of System Xc^- responsible for the transport of cystine from the extracellular to the intracellular. The nuclear factor erythroid-like 2 (Nrf2) and transcription factor 4 (activating factor 4, ATF4) can induce SLC7A11 expression when cells are in a state of oxidative stress and L-cysteine deficiency [45]. Studies have found that SLC7A11 is highly expressed in tumor tissues, and that high expression of SLC7A11 can inhibit ROS-induced ferroptosis. p53 leads to cystine deficiency by inhibiting the expression of SLC7A11, which in turn increases the sensitivity to ferroptosis. Importantly, the survival and proliferation of SLC7A11-overexpressing cancer cells are dependent on glucose, such tumors may be sensitive to glucose-blocking drugs, also suggesting a role for SLC7A11 in modulating nutrient dependence and demonstrating another therapeutic strategy for tumors with high SLC7A11 expression [46].

2.6.2 p53 regulates ferroptosis

p53 is a widely recognized tumor suppressor that can induce senescence and programmed cell death in human cancer. It can affect ferroptosis of tumor cells through transcriptional or posttranslational mechanisms. Also, it can inhibit tumors by regulating cell cycle arrest, apoptosis, or premature aging.

Increasing the stability of wild-type p53 can promote the expression of its transcriptional target gene CDKN1A (encoding p21 protein), which can increase the intracellular glutathione level, inhibit the accumulation of ROS, and negatively regulate ferroptosis of cancer cells. Acetylation-deficient p53 increased the sensitivity of tumor cells to ferroptosis by inhibiting *SLC7A11* expression and System Xc^- function.

Jiang et al. [47] found that three lysines in the DNA-binding domain of p53 were mutated to arginine (K117/161/162R, namely $p53^{3KR}$), $p53^{3KR}$ can further restrict cystine uptake by inhibiting SLC7A11 gene transcription, make tumor cells more sensitive to oxidative stress-induced ferroptosis. Wang et al. [48] found that the site K98 in the DNA-binding domain of p53 is particularly important for the regulation of SLC7A11. Additionally, the mutant S47 of p53 fails to inhibit the transcription of SLC7A11 and induces ferroptosis resistance in hepatocellular carcinoma cells, increasing the risk of cancer in mice [49]. Moreover, p53 can make colon cancer cells insensitive to ferroptosis by inhibiting dipeptidyl peptidase 4 activity [50].

2.6.3 NRF2 regulates ferroptosis

NRF2 is an important transcriptional regulator in oxidative reactions [51], and its overexpression can inhibit apoptosis and lead to drug resistance in some tumors [52]. NRF2 plays an important role in protecting hepatocellular carcinoma cells from ferroptosis [22]. After treatment of hepatocellular carcinoma cells with Erastin and Sorafenib, p62 inhibits the degradation of NRF2 and induces NRF2 accumulation in the nucleus through the inactivation of Keap1, thereby regulating downstream gene transcription. Inhibition of NRF2 by the alkaloid trigonelline can induce ferroptosis in hepatocellular carcinoma cells, and combined use with chemotherapeutic drugs has the application prospect of overcoming tumor drug resistance [22]. Thus, activation of the p62-Keap1-NRF2 pathway can activate ferroptosis to reverse tumor chemotherapeutic drug resistance.

2.6.4 ACSL4 regulates ferroptosis

ACSL4 is expressed on the mitochondrial outer membrane and endoplasmic reticulum and can convert long-chain fatty acids into fatty acyl-CoA, which plays an important role in lipid biosynthesis and fatty acid degradation. ACSL4 increases the sensitivity of cells to ferroptosis by accumulating long-chain polyunsaturated ω-6 fatty acids in the cell membrane [10]. Studies have shown that in basal-like breast cancer cell lines, liver cancer cells, leukemia cells, and prostate cancer cells, the expression level of ACSL4 can be used to predict the sensitivity of tumor cells to ferroptosis [10, 28, 52]. The results suggest that ACSL4 is expected to be a potential target and biological marker for targeting ferroptosis in tumor therapy.

2.6.5 GPX4

Glutathione peroxidase 4 (GPX4) is the only glutathione peroxidase that can use glutathione as the electron donor to reduce the toxic lipid hydroperoxides in biofilms to corresponding alcohols. Tumor cells with high GPX4 show impaired proliferation, decreased proliferation, and inhibition of angiogenesis. Overexpression of GPX4 in hepatocellular carcinoma cells can inhibit the formation and development of hepatocellular carcinoma by decreasing ROS level, increasing glutathione and decreasing the formation of the cytokine-cytokine IL-8, inhibiting cell cycle progression and cell migration [53]. Based on the clinicopathological study and in vitro cell death analysis, it was found that overexposure to GPX4 in diffuse large B-cell lymphoma (DLBCL) inhibited ROS-induced ferroptosis [54]. GPX4 is a major target molecule for ferroptosis inducers such as erastin and RSL3. Erastin inhibits GPX4 activity by depleting glutathione, whereas RSL3 can directly inhibit GPX4 activity. In addition, previous studies have demonstrated that GPX4 can induce ferroptosis in mouse tumor xenograft models [33].

2.6.6 FSP1 inhibits ferroptosis

FSP1 inhibits ferroptosis FSP1 was originally named mitochondrial apoptosis-inducing factor 2 (AIFM2), as the newly discovered GPX4-independent ferroptosis inhibitor, and its expression is closely related to the sensitivity of tumor cells to ferroptosis. Recently, Doll's group and Bersuker's group simultaneously screened independently and found that FSP1 levels are different in different cell lines, and the resistance level of various tumor cell lines to ferroptosis was positively correlated with the FSP1 level, which results in differences in the sensitivity of different tumor cell lines to ferroptosis [18, 19]. Additionally, it was also reported that FSP1 was a novel KEAP1/NRF2 target gene regulating ferroptosis and radioresistance in lung cancers [55]. These achievements provide important evidences for the development of drug-targeted ferroptosis in tumors.

3. Ferroptosis and neurodegenerative diseases

Iron homeostasis is critical for brain and neural development and cognitive function, especially in the fetal or early neonatal period, iron deficiency can severely affect neurodevelopment, leading to impaired memory and learning [56]. Iron accumulates gradually in the brain with age, and accumulation studies have shown that iron accumulation is related to neurodegenerative diseases such as Alzheimer's disease, Parkinson's disease, and amyotrophic lateral sclerosis [57]. In recent years, studies have found that there are main characteristics of ferroptosis such as increased lipid peroxidation, decreased glutathione, and GPX4 inhibition in neurodegenerative diseases and cognitive impairment. The use of ferroptosis inhibitors can effectively protect neurons and improve cognitive function. Related research progress is shown in **Table 2**.

3.1 Ferroptosis and Alzheimer's disease

Patients with Alzheimer's disease possess destabilization of metal metabolism, inflammatory response, oxidative stress, abnormal mitochondrial function, and impaired glial function [57, 58]. Studies have shown that iron accumulation in the

Diseases	Related studies	References
Alzheimer's disease	Overexpression or increase of phosphorylated tau protein can induce neuronal ferroptosis, and α-lipoic acid can inhibit tau protein-induced ferroptosis	[63]
	Hippocampal neuronal death and cognitive decline in brain *Gpx4*-induced knockout mice	[68]
	Ferritin levels in cerebrospinal fluid can predict the progression of Alzheimer's disease	[73]
Parkinson's disease	Activation of protein kinase C triggers ferroptosis	[79]
	Serine/threonine protein kinase is involved in Erastin-induced ferroptosis	[85]
	Astrocytes provide neurons with GSTM2 to protect neurons from oxidative damage	[81–83]
Amyotrophic lateral sclerosis	Neuronal *Gpx4*-inducible knockout mice develop symptoms of amyotrophic lateral sclerosis	[68]

Gpx4: glutathione peroxidase 4; GSTM2:Glutathione S-transferase Mu2.

Table 2.
Ferroptosis in neurodegenerative diseases.

brain is associated with the formation of senile plaques and neurofibrillary tangles, elevated iron levels in the brain increase the risk of Alzheimer's disease, and ferritin levels in cerebrospinal fluid predict the progression from mild cognitive impairment to Alzheimer's disease [59–61]. The chronic inflammation, neuronal degeneration, and lack of downstream apoptosis indicators associated with Alzheimer's disease suggest the existence of other cell death manners such as ferroptosis in Alzheimer's disease [62–64].

An investigation on the Gpx4-specific knockout mice in cerebral cortex and hippocampal neurons exhibited cognitive decline and degeneration of hippocampal neurons in the water maze test, after feeding with a vitamin-E-rich diet or Lip-1, the neuronal degeneration of the mice was significantly alleviated, suggesting that ferroptosis plays an important role in neuronal degeneration [62]. Another study found that overexpression or hyperphosphorylation of tau protein can induce ferroptosis in neurons, while α-lipoic acid can rescue neurons by downregulating TfR1, reducing p38 phosphorylation level, and upregulating the expression of Slc7a11 and Gpx4 [63]. In addition, feeding with a deuterated polyunsaturated fatty acid in a mouse model of Alzheimer's disease can alleviate the lipid peroxidation of tissues and reduces β-amyloid deposition [64, 65].

3.2 Ferroptosis and Parkinson's disease

An important feature of Parkinson's disease is iron accumulation in neurons and substantia nigra glia, and the concentration of iron accumulation is positively correlated with disease severity [66, 67]. Significant changes in iron regulatory protein 1(IRP1), divalent metal transporter 1(DMT1), and other key proteins involved in iron homeostasis have been observed in Parkinson's disease patients and mouse models [68–72]. The τ knockout mice developed parkinsonism with iron accumulation in the nigra, which can be inhibited by iron chelators [73–75]. In addition to elevated iron levels in the substantia nigra pars compactus, parkinsonism is also characterized by ferroptosis, such as reduced glutathione depletion and lipid peroxidation [76], iron chelators and N-acetylcystine can alleviate and improve some of the symptoms in patients and mouse model of Parkinson's disease [77, 78], suggesting that ferroptosis may be involved in the occurrence and development of Parkinson's disease.

Do Van et al. [79] found that dopaminergic neurons' ferroptosis occurred in LUHMES cell lines, brain tissue slices cultured *in vitro*, and 1-methyl-4-phenyl-1,2,3,6-tetrahydropyridine (MPTP)-induced Parkinson's disease models. However, the use of Fer-1, Lip-1, and iron chelators can alleviate or reverse the symptoms of Parkinson's disease. Gouel et al. [80] found that human platelet lysate can make LUHMES cells resistant to Earstin-induced neuronal ferroptosis. In addition, astrocytes have a strong iron storage capacity, which prevent iron overload in neurons [81]. Astrocytes provide neurons with glutathione S-transferase Mu2 (GSTM2) and other antioxidant factors to protect neurons from oxidative damage. In conclusion, the dysregulation of the interaction between astrocytes and neurons may lead to ferroptosis in dopaminergic neurons [82, 83].

3.3 Ferroptosis and amyotrophic lateral sclerosis

Iron accumulation occurred in the brain of amyotrophic lateral sclerosis model mice [84–86], and the therapeutic effect of iron chelators confirms the role of iron in the pathogenesis of amyotrophic lateral sclerosis. Patients with amyotrophic lateral sclerosis have increased lipid peroxidation in cerebrospinal fluid and plasma and decreased levels of reduced glutathione in the motor cortex, suggesting the possibility of ferroptosis occurrence [87, 88]. Knockout of Gpx4 in mouse neurons can cause symptoms of amyotrophic lateral sclerosis, mainly characterized by rapid paralysis, severe muscle atrophy, and death, which is related to ferroptosis of spinal motor neurons [62]. However, no significant neurodegeneration was observed in the cortex of neuronal Gpx4-inducible knockout and other Gpx4-selective cortical neuron knockout mouse models. It is suggested that Gpx4 plays an important role in the process of ferroptosis of spinal motor neurons [62].

4. Ferroptosis and cardiovascular disease

Cardiomyocyte and neuron death in cardiovascular and cerebrovascular diseases are related to a variety of cell death manners, and ferroptosis is also involved. The relevant research progress is shown in **Table 3**.

4.1 Ferroptosis and cardiovascular disease

In some pathological conditions, the heart exhibits excessive accumulation of iron, production of reactive oxygen species, and pathological transformation of membrane lipids, which are all important factors that constitute ferroptosis. So far, there are few studies directly linking ferroptosis with cardiovascular disease. The latest research results of the Wang group in 2019 revealed the important role of ferroptosis in cardiomyopathy and ischemia–reperfusion-induced cardiac injury for the first time [89]. This landmark discovery provides a new strategy for the prevention and treatment of cardiomyopathy and other heart diseases.

Diseases	Related studies	References
ischemia–reperfusion	In an isolated mouse cardiac ischemia–reperfusion model, inhibition of glutamine metabolism can attenuate ferroptosis-induced cardiac injury	[93]
	ferroptosis inhibitors and iron chelators can effectively alleviate myocardial injury induced by cardiac ischemia–reperfusion in mice	[89]
Doxorubicin-induced myocardial injury	Doxorubicin induces ferroptosis in cardiomyocytes via heme oxygenase; iron accumulation and lipid peroxidation mainly occur in mitochondria	[89]
	Heme oxygenase inhibitors, ferroptosis inhibitors, mitochondrial antioxidant inhibitors, iron chelators, etc., can effectively reverse doxorubicin-induced myocardial injury	[89]
Myocardial damage after heart transplantation	Ferroptosis regulates neutrophil recruitment after cardiac transplantation in mice	[105]
Ischemic stroke	Hypoxia-inducible factor prolyl hydroxylase may be a potential target of iron chelators to inhibit neuronal ferroptosis	[115]
	Inhibition of ferroptosis protects neurons in mice with middle cerebral artery occlusion, and the interaction between iron and tau protein is pleiotropically regulated	[105]
Hemorrhagic stroke	(−)-Epicatechin alleviates early brain injury in hemorrhagic stroke by reducing cerebral iron accumulation and ferroptosis-related protein expression	[116]
	Ferroptosis inhibitors attenuate neuronal death in brain slices and in a mouse model of hemorrhagic stroke	[117–118]
	Increased glutathione peroxidase 4 expression can avoid neuronal ferroptosis and improve prognosis	[119]

Table 3.
Ferroptosis in cardiovascular and cerebrovascular diseases.

4.1.1 Ferroptosis is involved in tissue and organ induced by ischemia: reperfusion damage

During cardiac ischemia–reperfusion, excess reactive oxygen species, lipid peroxidation, and iron accumulation caused by the release of iron in heme will be produced [90–92]. Gao et al. [93] established an isolated mouse cardiac ischemia–reperfusion model and found that inhibiting glutamine metabolism could inhibit ferroptosis, thereby reducing cardiac injury. Fang et al. [89] established an in vivo myocardial ischemia–reperfusion model and found that Fer-1 and iron chelators can significantly reduce the acute and chronic cardiac injury of ischemia–reperfusion, confirming the role of ferroptosis in cardiac ischemia–reperfusion injury. In addition, ferroptosis is also involved in ischemia–reperfusion injury in the kidney [94] and liver [95].

4.1.2 Ferroptosis is involved in antitumor drug-induced myocardial injury

As a broad-spectrum antitumor drug, adriamycin was limited in clinical use due to its cardiotoxicity. Autophagy, apoptosis, necrosis, and other cell death type are involved in the myocardial injury caused by adriamycin [96–98]. Fang et al. [89] found that ferroptosis occurred in cardiomyopathy induced by doxorubicin in mice deficient in apoptosis and proposed that heme oxygenase 1 (HO-1) may be a key regulator in this procedure. They also found that iron accumulation and lipid peroxidation in cardiomyocytes occur in mitochondria and the mitochondria-targeting antioxidant MitoTEMPO can effectively inhibit ferroptosis and protect the heart.

4.1.3 Ferroptosis is involved in myocardial injury after heart transplantation

In Li's studies [99], it was found that the recruitment of neutrophils after heart transplantation is regulated by ferroptosis. The donor heart can induce ferroptosis in cardiomyocytes due to ischemia, hypoxia, and other reasons after transplantation, and the cellular contents are released and recruit neutrophils to produce necrotic inflachannelled by TLR4 / Trif/type I interflammatory via TLR/Trif/type I interferon pathway. Fer-1 can reduce arachidyl phosphatidylethanolamine after heart transplantation and decreased cardiomyocyte death and neutrophil recruitment.

4.1.4 Ferroptosis and diabetic cardiomyopathy

In diabetes, persistent high blood glucose and insulin resistance can cause a vicious circle by altering cellular metabolism, promoting the accumulation of peroxidation and the death of cells. So far, diabetes has been verified to be associated with abnormal iron metabolism. For example, systemic iron overload can contribute to abnormal glucose metabolism and the onset of type 2 diabetes (T2DM) [100] and aggravate insulin resistance [101]. Recently, Cai group [102] identified the role of ferroptosis in DCM and reported that Nrf2 activation by sulforaphane inhibited ferroptosis and prevented DCM, suggesting that it is feasible to treat DCM by inhibiting ferroptosis. Due to the limited regenerative capacity of the myocardium in mammalian adult hearts, inhibition of cardiomyocyte death might be one of the important ways to alleviate DCM [103]. In our studies, we induced DCM models in diabetic C57BL6 mice and treated with canagliflozin and found that canagliflozin mitigates ferroptosis and improves myocardial oxidative stress in mice with diabetic cardiomyopathy [104]. Taken together, taking ferroptosis as the starting point may provide a new strategy for the prevention and control of DCM.

4.2 Ferroptosis and cerebrovascular disease

Both ischemic stroke and hemorrhagic stroke can lead to neuronal ferroptosis [105, 106].

4.2.1 Ferroptosis and ischemic stroke

Before the discovery of ferroptosis, iron accumulation in clinical and animal models of ischemic stroke has been shown to aggravate neuronal damage during reperfusion [107–110]. Iron chelators can reduce the risk of post-ischemic stroke in experimental animals [111–114]. Speer et al. [115] proposed that ferroptosis leads to neuronal death after cerebral ischemia, and hypoxia-inducible factor prolyl hydroxylase may be the target for the beneficial effects of iron chelators. Inhibition of ferroptosis in a mouse model can protect neurons from ischemia–reperfusion injury [105].

4.2.2 Ferroptosis and hemorrhagic stroke

Chang et al. [116] found that epicatechin could alleviate early brain injury in hemorrhagic stroke by reducing cerebral iron accumulation and ferroptosis-related protein expression. Later, they found that Fer-1 could alleviate hemoglobin-induced brain injury. Cell death in slices and alleviation of neuronal death in a mouse model of collagenase-induced hemorrhagic stroke [117]. At the same time, Zille et al. [118] found that ferroptosis inhibitors such as Fer-1 and deferoxamine can inhibit the production of ferroptosis in mice. The expression level of Gpx4 in rats with acute hemorrhagic stroke decreased sharply, and increasing the level of Gpx4 could avoid secondary ferroptosis injury in neurons and improve the prognosis of hemorrhagic stroke [119]. Therefore, the ferroptosis pathway may be involved in the process of neuronal death in stroke, and it is speculated that targeted inhibition of ferroptosis may be an effective treatment for alleviating stroke.

5. Conclusion

Besides tumors, neurodegenerative diseases, cardiovascular and cerebrovascular diseases, ferroptosis has also been reported in liver diseases such as non-alcoholic fatty liver disease and non-alcoholic steatohepatitis [120–124].

ROS-induced ferroptosis can inhibit tumor growth and increase the sensitivity of tumor cells to chemotherapy and radiotherapy. Contrary to tumor treatment strategies, ferroptosis can promote the occurrence and development of neurodegenerative diseases and cardiovascular and cerebrovascular diseases. Therefore, relevant translational medicine research mainly focuses on the discovery of small molecules that can effectively inhibit ferroptosis. These small-molecule activators targeting ferroptosis can be used directly as chemotherapeutics, or as chemosensitizers in combination with chemotherapeutics. However, ferroptosis is complex in different types of tumors and different gene mutations (such as p53 or RAS mutations), and its feasibility in preclinical and clinical research needs to be further studied. Notably, the discovery of GPX4 pathway-independent FSP1 and the discovery of new mechanisms and targets such as CD8+ T cells inducing ferroptosis in tumor cells through the release of interferon-gamma [18, 19, 125], it provides a new perspectives and strategy for tumor treatment and drug discovery.

Iron accumulation and ferroptosis in the brain and nerve tissue have been proved to be closely related to Alzheimer's disease and Parkinson's disease. There is a direct relationship between the pathogenesis of various neurodegenerative diseases such as Parkinson's disease and amyotrophic lateral sclerosis. At present, various clinical trials using iron chelators to treat neurodegenerative diseases are emerging, but there is still no effective treatment for stroke. Given the important role of ferroptosis in neuronal death after stroke, effective inhibition of ferroptosis is expected to provide a new strategy for preventing neuronal death caused by stroke.

Similar to the pathogenesis of neurodegenerative diseases, many cardiac diseases share common ferroptosis features, such as iron overload, oxidative stress, endoplasmic reticulum stress, and mitochondrial dysfunction. Previous studies by the author's team suggest that ferroptosis inhibitors can effectively prevent and treat cardiomyopathy and heart failure induced by myocardial cell iron overload, doxorubicin-induced cardiotoxicity, and cardiac ischemia–reperfusion [89]. Five different approaches, including ferroptosis inhibitors, iron chelators, mitochondria-specific antioxidants, heme oxygenase 1 inhibitors, and low-iron diets, can effectively prevent ferroptosis in cardiomyocytes, thereby protecting the heart. And these ferroptosis inhibitors are relatively safe and feasible in mice. It provides an optimistic prospect for clinical translational research on targeting ferroptosis to prevent and treat heart disease [121, 126, 127].

With a view to clinical translation, here are some issues need to be considered, e.g., which disease or tumor needs to be considered for ferroptosis-targeted therapies? In clinical or preclinical experiments, drugs targeting ferroptosis need high tissue-organ specificity and fewer adverse reactions, and nano-targeted drug delivery systems have shown some advantages [128, 129]. Although there is a growing awareness of ferroptosis, some key scientific questions related to ferroptosis still need to be resolved, such as what are the key executive molecules in ferroptosis? To what extent is lipid peroxidation related to ferroptosis? Does ferroptosis exist in physiological processes? Is ferroptosis conservative in the evolutionary process? We are well aware of the long road ahead. With the deepening and expansion of ferroptosis-related research, we believe that it will provide a basis for the clinical translation of targeting ferroptosis to prevent and treat major chronic diseases.

Acknowledgements

This work was supported by grants from Medicine and Health Science and Technology Plan Projects of Zhejiang Province (YS, 2020PY029), Science and Technology Innovation Special Project of Jiaxing Science and Technology Bureau (YS, 2020AY30003), Zhejiang Provincial Health Science and Technology Program of Traditional Chinese Medicine (YS, 2021ZB283), Zhejiang Basic Public Welfare Research Program (YS, LGF18H200004), and Jiaxing Key Laboratory of Diabetic Angiopathy.

Conflict of interest

The authors declare no conflict of interest.

Author details

Yanbo Shi[1*], Junyong Zhang[2], Kaitao Luo[3], Sunfeng Pan[4], Hanqiang Shi[5], Lie Xiong[5] and Shuqin Du[6]

1 Central Laboratory of Molecular Medicine Center, Zhejiang Chinese Medical University Affiliated Jiaxing Traditional Chinese Medicine Hospital, Jiaxing Key Laboratory of Diabetic Angiopathy Research, Zhejiang, China

2 College of Biological, Chemical Sciences and Engineering, Jiaxing University, Zhejiang, China

3 Jiaxing Key Laboratory of Integrative Rehabilitation of Cerebrovascular Disease, Zhejiang Chinese Medical University Affiliated Jiaxing Traditional Chinese Medicine Hospital, Zhejiang, China

4 Department of Burns and Plastic Surgery, Zhejiang Chinese Medical University Affiliated Jiaxing Traditional Chinese Medicine Hospital, Jiaxing Key Laboratory of Diabetic Angiopathy Research, Jiaxing Burn and Wound Repair Therapy Center, Zhejiang, China

5 Central Laboratory of Molecular Medicine Center, Jiaxing Traditional Chinese Medicine Hospital, Jiaxing Key Laboratory of Diabetic Angiopathy Research, Jiaxing, Zhejiang, China

6 School of Pharmacy, Zhejiang University of Technology, Hangzhou, China

*Address all correspondence to: shiyanbocas@163.com

IntechOpen

© 2022 The Author(s). Licensee IntechOpen. This chapter is distributed under the terms of the Creative Commons Attribution License (http://creativecommons.org/licenses/by/3.0), which permits unrestricted use, distribution, and reproduction in any medium, provided the original work is properly cited.

References

[1] World Health Statistics. Monitoring Health for the SDGs, Sustainable Development Goals. Genève: World Health Organization; 2021

[2] Hentze MW, Muckenthaler MU, Andrews NC. Balancing acts: Molecular control of mammalian iron metabolism. Cell. 2004;**117**:285-297

[3] Rouault TA. Cell biology. An ancient gauge for iron. Science. 2009;**326**(5953):676-677

[4] Dixon SJ, Lemberg KM, Lamprecht MR, Skouta R, Zaitsev EM, Gleason CE, et al. Ferroptosis: An iron-dependent form of nonapoptotic cell death. Cell. 2012;**149**(5):1060-1072

[5] Cao JY, Dixon SJ. Mechanisms of ferroptosis. Cellular and Molecular Life Sciences. 2016;**73**(11-12):2195-2209

[6] Dixon SJ, Stockwell BR. The role of iron and reactive oxygen species in cell death. Nature Chemical Biology. 2014;**10**(1):9-17

[7] Kagan VE, Mao G, Qu F, et al. Oxidized arachidonic and adrenic PEs navigate cells to ferroptosis. Nature Chemical Biology. 2017;**13**(1):81-90

[8] Mancias JD, Wang X, Gygi SP, Harper JW, Kimmelman AC. Quantitative proteomics identifies NCOA4 as the cargo receptor mediating ferritinophagy. Nature. 2014;**509**(7498):105-1090

[9] Wenzel SE, Tyurina YY, Zhao J, St Croix CM, Dar HH, Mao G, et al. PEBP1 wardens ferroptosis by enabling lipoxygenase generation of lipid death signals. Cell. 2017;**171**(3):628-641

[10] Doll S, Proneth B, Tyurina YY, Panzilius E, Kobayashi S, Ingold I, et al. ACSL4 dictates ferroptosis sensitivity by shaping cellular lipid composition. Nature Chemical Biology. 2017;**13**(1):91-98

[11] Hangauer MJ, Viswanathan VS, Ryan MJ, Bole D, Eaton JK, Matov A, et al. Drug-tolerant persister cancer cells are vulnerable to GPX4 inhibition. Nature. 2017;**9**(7679):247-250

[12] Dixon SJ, Winter GE, Musavi LS, Lee ED, Snijder B, Rebsamen M, et al. Human haploid cell genetics reveals roles for lipid metabolism genes in nonapoptotic cell death. ACS Chemical Biology. 2015;**10**(7):1604-1609

[13] Gao M, Monian P, Pan Q, Zhang W, Xiang J, Jiang X. Ferroptosis is an autophagic cell death process. Cell Research. 2016;**26**(9):1021-1032

[14] Friedmann Angeli JP, Conrad M. Selenium and GPX4, a vital symbiosis. Free Radical Biology & Medicine. 2018;**1**(127):153-159

[15] Maiorino M, Conrad M, Ursini F. GPx4, lipid peroxidation, and cell death: Discoveries, rediscoveries, and open issues. Antioxidants & Redox Signaling. 2018;**29**(1):61-74

[16] Lu L, Hope BT, Shaham Y. The cystine-glutamate transporter in the accumbens: A novel role in cocaine relapse. Trends in Neurosciences. 2004;**27**(2):74-76

[17] Wang H, An P, Xie E, Wu Q, Fang X, Gao H, et al. Characterization of ferroptosis in murine models of hemochromatosis. Hepatology. 2017;**66**(2):449-465

[18] Bersuker K, Hendricks JM, Li Z, Magtanong L, Ford B, Tang PH, et al.

The CoQ oxidoreductase FSP1 acts parallel to GPX4 to inhibit ferroptosis. Nature. 2019;**575**(7784):688-692

[19] Doll S, Freitas FP, Shah R, Aldrovandi M, da Silva MC, Ingold I, et al. FSP1 is a glutathione-independent ferroptosis suppressor. Nature. 2019;**575**(7784):693-698

[20] Stockwell BR, Friedmann Angeli JP, Bayir H, Bush AI, Conrad M, Dixon SJ, et al. Ferroptosis: A regulated cell death Nexus linking metabolism, redox biology, and disease. Cell. 2017;**171**(2):273-285

[21] Louandre C, Marcq I, Bouhlal H, Lachaier E, Godin C, Saidak Z, et al. The retinoblastoma (Rb) protein regulates ferroptosis induced by sorafenib in human hepatocellular carcinoma cells. Cancer Letters. 2015;**356**(2):971-977

[22] Sun X, Ou Z, Chen R, Niu X, Chen D, Kang R, et al. Activation of the p62-Keap1-NRF2 pathway protects against ferroptosis in hepatocellular carcinoma cells. Hepatology. 2016;**63**(1):173-184

[23] Harrison PM, Arosio P. The ferritins: Molecular properties, iron storage function and cellular regulation. Biochimica et Biophysica Acta. 1996;**1275**(3):161-203

[24] Raghunath A, Sundarraj K, Arfuso F, Sethi G, Perumal E. Dysregulation of Nrf2 in Hepatocellular Carcinoma: Role in Cancer Progression and Chemoresistance. Cancers (Basel). 3 Dec 2018;**10**(12):481. DOI: 10.3390/cancers10120481. PMID: 30513925; PMCID: PMC6315366

[25] Arlt A, Sebens S, Krebs S, Geismann C, Grossmann M, Kruse ML, et al. Inhibition of the Nrf2 transcription factor by the alkaloid trigonelline renders pancreatic cancer cells more susceptible to apoptosis through decreased proteasomal gene expression and proteasome activity. Oncogene. 2013;**32**(40):4825-4835

[26] Suzuki T, Motohashi H, Yamamoto M. Toward clinical application of the Keap1-Nrf2 pathway. Trends in Pharmacological Sciences. 2013;**34**(6):340-346

[27] Sun X, Niu X, Chen R, He W, Chen D, Kang R, et al. Metallothionein-1G facilitates sorafenib resistance through inhibition of ferroptosis. Hepatology. 2016;**64**(2):488-500

[28] Chen WC, Wang CY, Hung YH, Weng TY, Yen MC, Lai MD. Systematic analysis of gene expression alterations and clinical outcomes for long-chain acyl-coenzyme a synthetase family in Cancer. PLoS One. 2016;**11**(5):e0155660

[29] Kinoshita H, Okabe H, Beppu T, Chikamoto A, Hayashi H, Imai K, et al. Cystine/glutamic acid transporter is a novel marker for predicting poor survival in patients with hepatocellular carcinoma. Oncology Reports. 2013;**29**(2):685-689

[30] Efferth T, Dunstan H, Sauerbrey A, Miyachi H, Chitambar CR. The anti-malarial artesunate is also active against cancer. International Journal of Oncology. 2001;**18**(4):767-773

[31] Eling N, Reuter L, Hazin J, Hamacher-Brady A, Brady NR. Identification of artesunate as a specific activator of ferroptosis in pancreatic cancer cells. Oncoscience. 2015;**2**(5):517-532

[32] Yamaguchi Y, Kasukabe T, Kumakura S. Piperlongumine rapidly induces the death of human pancreatic cancer cells mainly through the induction of ferroptosis. International Journal of Oncology. 2018;**52**(3):1011-1022

[33] Yang WS, SriRamaratnam R, Welsch ME, Shimada K, Skouta R,

Viswanathan VS, et al. Regulation of ferroptotic cancer cell death by GPX4. Cell. 2014;**156**(1-2):317-331

[34] Wu J, Minikes AM, Gao M, Bian H, Li Y, Stockwell BR, et al. Publisher correction: Intercellular interaction dictates cancer cell ferroptosis via NF2-YAP signalling. Nature. 2019;**572**(7770):E20

[35] Yang WH, Ding CC, Sun T, Rupprecht G, Lin CC, Hsu D, et al. The hippo pathway effector TAZ regulates ferroptosis in renal cell carcinoma. Cell Reports. 2019;**28**(10):2501-2508

[36] Ma S, Henson ES, Chen Y, Gibson SB. Ferroptosis is induced following siramesine and lapatinib treatment of breast cancer cells. Cell Death & Disease. 2016;**7**(7):e2307

[37] Hasegawa M, Takahashi H, Rajabi H, Alam M, Suzuki Y, Yin L, et al. Functional interactions of the cystine/glutamate antiporter, CD44v and MUC1-C oncoprotein in triple-negative breast cancer cells. Oncotarget. 2016;**7**(11):11756-11769

[38] Timmerman LA, Holton T, Yuneva M, Louie RJ, Padró M, Daemen A, et al. Glutamine sensitivity analysis identifies the xCT antiporter as a common triple-negative breast tumor therapeutic target. Cancer Cell. 2013;**24**(4):450-465

[39] Lanzardo S, Conti L, Rooke R, Ruiu R, Accart N, Bolli E, et al. Immunotargeting of antigen xCT attenuates stem-like cell behavior and metastatic progression in breast cancer. Cancer Research. 2016;**76**(1):62-72

[40] Habashy HO, Powe DG, Staka CM, Rakha EA, Ball G, Green AR, et al. Transferrin receptor (CD71) is a marker of poor prognosis in breast cancer and can predict response to tamoxifen. Breast Cancer Research and Treatment. 2010;**119**(2):283-293

[41] Tonik SE, Shindelman JE, Sussman HH. Transferrin receptor is inversely correlated with estrogen receptor in breast cancer. Breast Cancer Research and Treatment. 1986;**7**(2):71-76

[42] Martin-Sanchez D, Fontecha-Barriuso M, Sanchez-Niño MD, Ramos AM, Cabello R, Gonzalez-Enguita C, et al. Cell death-based approaches in treatment of the urinary tract-associated diseases: A fight for survival in the killing fields. Cell Death & Disease. 2018;**9**(2):118

[43] Mazdak H, Yazdekhasti F, Movahedian A, Mirkheshti N, Shafieian M. The comparative study of serum iron, copper, and zinc levels between bladder cancer patients and a control group. International Urology and Nephrology. 2010;**42**(1):89-93

[44] Tang M, Chen Z, Wu D, Chen L. Ferritinophagy/ferroptosis: Iron-related newcomers in human diseases. Journal of Cellular Physiology. 2018;**233**(12):9179-9190

[45] Koppula P, Zhang Y, Zhuang L, Gan B. Amino acid transporter SLC7A11/xCT at the crossroads of regulating redox homeostasis and nutrient dependency of cancer. Cancer Communication (Lond). 2018;**38**(1):12

[46] Koppula P, Zhang Y, Shi J, Li W, Gan B. The glutamate/cystine antiporter SLC7A11/xCT enhances cancer cell dependency on glucose by exporting glutamate. Journal of Biology Chemistry. 2017;**292**(34):14240-14249

[47] Jiang L, Kon N, Li T, Wang SJ, Su T, Hibshoosh H, et al. Ferroptosis as a p53-mediated activity during

tumour suppression. Nature. 2015;**520**(7545):57-62

[48] Wang SJ, Li D, Ou Y, Jiang L, Chen Y, Zhao Y, et al. Acetylation is crucial for p53-mediated ferroptosis and tumor suppression. Cell Reports. 2016;**17**(2):366-373

[49] Jennis M, Kung CP, Basu S, Budina-Kolomets A, Leu JI, Khaku S, et al. An African-specific polymorphism in the TP53 gene impairs p53 tumor suppressor function in a mouse model. Genes Development. 2016;**30**(8):918-930

[50] Xie Y, Zhu S, Song X, Sun X, Fan Y, Liu J, et al. The tumor suppressor p53 limits ferroptosis by blocking DPP4 activity. Cell Reports. 2017;**20**(7):1692-1704

[51] Ma Q. Role of nrf2 in oxidative stress and toxicity. Annual Review of Pharmacology and Toxicology. 2013;**53**:401-426

[52] Yuan H, Li X, Zhang X, Kang R, Tang D. Identification of ACSL4 as a biomarker and contributor of ferroptosis. Biochemical and Biophysical Research Communications. 2016;**478**(3):1338-1343

[53] Rohr-Udilova N, Bauer E, Timelthaler G, et al. Impact of glutathione peroxidase 4 on cell proliferation, angiogenesis and cytokine production in hepatocellular carcinoma. Oncotarget, 2018;**9**: 10054-10068.

[54] Kinowaki Y, Kurata M, Ishibashi S, Ikeda M, Tatsuzawa A, Yamamoto M, et al. Glutathione peroxidase 4 overexpression inhibits ROS-induced cell death in diffuse large B-cell lymphoma. Laboratory Investigation. 2018;**98**(5):609-619

[55] Emmanuel N, Li H, Chen J, Zhang Y. FSP1, a novel KEAP1/NRF2 target gene regulating ferroptosis and radioresistance in lung cancers. Oncotarget. 19 Oct 2022;**13**:1136-1139. DOI: 10.18632/oncotarget.28301. PMID: 36264074; PMCID: PMC9584440

[56] Radlowski EC, Johnson RW. Perinatal iron deficiency and neurocognitive development. Frontiers in Human Neuroscience. 2013;**23**(7):585

[57] Belaidi AA, Bush AI. Iron neurochemistry in Alzheimer's disease and Parkinson's disease: Targets for therapeutics. Journal of Neurochemistry. 2016;**139**(Suppl. 1):179-197

[58] Bush AI, Curtain CC. Twenty years of metallo-neurobiology: Where to now? European Biophysics Journal. 2008;**37**(3):241-245

[59] Ayton S, Faux NG, Bush AI. Alzheimer’s disease neuroimaging initiative. Ferritin levels in the cerebrospinal fluid predict Alzheimer's disease outcomes and are regulated by APOE. Nature Communications. 2015;**19**(6):6760

[60] Connor JR, Snyder BS, Beard JL, Fine RE, Mufson EJ. Regional distribution of iron and iron-regulatory proteins in the brain in aging and Alzheimer's disease. Journal of Neuroscience Research. 1992;**31**(2):327-335

[61] Bilgic B, Pfefferbaum A, Rohlfing T, Sullivan EV, Adalsteinsson E. MRI estimates of brain iron concentration in normal aging using quantitative susceptibility mapping. Neuroimage. 2012;**59**(3):2625-2635

[62] Hambright WS, Fonseca RS, Chen L, Na R, Ran Q. Ablation of ferroptosis regulator glutathione peroxidase 4 in forebrain neurons promotes cognitive impairment and neurodegeneration. Redox Biology. 2017;**12**:8-17

[63] Zhang YH, Wang DW, Xu SF, Zhang S, Fan YG, Yang YY, et al. α-Lipoic acid improves abnormal behavior by mitigation of oxidative stress, inflammation, ferroptosis, and tauopathy in P301S Tau transgenic mice. Redox Biolology. 2018;**14**:535-548

[64] Raefsky SM, Furman R, Milne G, Pollock E, Axelsen P, Mattson MP, et al. Deuterated polyunsaturated fatty acids reduce brain lipid peroxidation and hippocampal amyloid β-peptide levels, without discernable behavioral effects in an APP/PS1 mutant transgenic mouse model of Alzheimer's disease. Neurobiology of Aging. 2018;**66**:165-176

[65] Yang WS, Kim KJ, Gaschler MM, Patel M, Shchepinov MS, Stockwell BR. Peroxidation of polyunsaturated fatty acids by lipoxygenases drives ferroptosis. Proceedings of the National Academy Science U S A. 2016;**113**:E4966-E4975

[66] Dexter DT, Wells FR, Agid F, Agid Y, Lees AJ, Jenner P, et al. Increased nigral iron content in postmortem parkinsonian brain. Lancet. 1987;**2**(8569):1219-1220

[67] Dexter DT, Wells FR, Lees AJ, Agid F, Agid Y, Jenner P, et al. Increased nigral iron content and alterations in other metal ions occurring in brain in Parkinson's disease. Journal of Neurochemistry. 1989;**52**(6):1830-1836

[68] Ayton S, Lei P, Duce JA, Wong BX, Sedjahtera A, Adlard PA, et al. Ceruloplasmin dysfunction and therapeutic potential for Parkinson disease. Annals of Neurology. 2013;**73**(4):554-559

[69] Khandelwal PJ, Herman AM, Moussa CE. Inflammation in the early stages of neurodegenerative pathology. Journal of Neuroimmunology. 2011;**238**(1-2):1-11

[70] Raina AK, Hochman A, Zhu X, Rottkamp CA, Nunomura A, Siedlak SL, et al. Abortive apoptosis in Alzheimer's disease. Acta Neuropathologica. 2001;**101**(4):305-310

[71] Boag MK, Roberts A, Uversky VN, Ma L, Richardson DR, Pountney DL. Ferritinophagy and α-Synuclein: Pharmacological Targeting of Autophagy to Restore Iron Regulation in Parkinson's Disease. International Journal of Molecular Sciences. 21 Feb 2022;**23**(4):2378. DOI: 10.3390/ijms23042378. PMID: 35216492; PMCID: PMC8878351

[72] Rao IY, Hanson LR, Johnson JC, Rosenbloom MH, Frey WH 2nd. Brain Glucose Hypometabolism and Iron Accumulation in Different Brain Regions in Alzheimer's and Parkinson's Diseases. Pharmaceuticals (Basel). 29 Apr 2022;**15**(5):551. DOI: 10.3390/ph15050551. PMID: 35631378; PMCID: PMC9143620

[73] Lei P, Ayton S, Finkelstein DI, Adlard PA, Masters CL, Bush AI. Tau protein: Relevance to Parkinson's disease. The International Journal of Biochemistry & Cell Biology. 2010;**42**(11):1775-1778

[74] Lei P, Ayton S, Moon S, Zhang Q, Volitakis I, Finkelstein DI, et al. Motor and cognitive deficits in aged tau knockout mice in two background strains. Molecular Neurodegeneration. 2014;**14**(9):29

[75] Lei P, Ayton S, Appukuttan AT, Volitakis I, Adlard PA, Finkelstein DI, et al. Clioquinol rescues parkinsonism and dementia phenotypes of the tau knockout mouse. Neurobiology of Disease. 2015;**81**:168-175

[76] Sian J, Dexter DT, Lees AJ, Daniel S, Agid Y, Javoy-Agid F, et al. Alterations in glutathione levels in Parkinson's disease and other neurodegenerative disorders

affecting basal ganglia. Annals of Neurology. 1994;**36**(3):348-355

[77] Farr AC, Xiong MP. Challenges and Opportunities of Deferoxamine Delivery for Treatment of Alzheimer's Disease, Parkinson's Disease, and Intracerebral Hemorrhage. Mol Pharm. 1 Feb 2021;**18**(2):593-609. DOI: 10.1021/acs.molpharmaceut.0c00474. Epub 2020 Oct 9. PMID: 32926630; PMCID: PMC8819678

[78] Monti DA, Zabrecky G, Kremens D, Liang TW, Wintering NA, Cai J, et al. N-acetyl cysteine may support dopamine neurons in Parkinson's disease: Preliminary clinical and cell line data. PLoS One. 2016;**11**(6):e0157602

[79] Do Van B, Gouel F, Jonneaux A, Timmerman K, Gelé P, Pétrault M, et al. Ferroptosis, a newly characterized form of cell death in Parkinson's disease that is regulated by PKC. Neurobiology of Disease. 2016;**94**:169-178

[80] Gouel F, Do Van B, Chou ML, Jonneaux A, Moreau C, Bordet R, et al. The protective effect of human platelet lysate in models of neurodegenerative disease: Involvement of the Akt and MEK pathways. Journal of Tissue Engineering and Regenerative Medicine. 2017;**11**:3236-3240

[81] Codazzi F, Pelizzoni I, Zacchetti D, Grohovaz F. Iron entry in neurons and astrocytes: A link with synaptic activity. Frontiers in Molecular Neuroscience. 2015;**3**(8):18

[82] Cui Z, Zhong Z, Yang Y, Wang B, Sun Y, Sun Q, et al. Ferrous Iron induces Nrf2 expression in mouse brain astrocytes to prevent neurotoxicity. Journal of Biochemical and Molecular Toxicology. 2016;**30**(8):396-403

[83] Ishii T, Warabi E, Mann GE. Circadian control of BDNF-mediated Nrf2 activation in astrocytes protects dopaminergic neurons from ferroptosis. Free Radical Biology & Medicine. 2019;**133**:169-178

[84] Gajowiak A, Styś A, Starzyński RR, Staroń R, Lipiński P. Misregulation of iron homeostasis in amyotrophic lateral sclerosis. Postępy Higieny i Medycyny Doświadczalnej (Online). 2016;**70**:709-721

[85] Moreau C, Danel V, Devedjian JC, Grolez G, Timmerman K, Laloux C, et al. Could conservative Iron chelation Lead to neuroprotection in amyotrophic lateral sclerosis? Antioxidants & Redox Signaling. 2018;**29**(8):742-748

[86] Veyrat-Durebex C, Corcia P, Mucha A, Benzimra S, Mallet C, Gendrot C, et al. Iron metabolism disturbance in a French cohort of ALS patients. Biomedical Research International. 2014;**2014**:485723

[87] Simpson EP, Henry YK, Henkel JS, Smith RG, Appel SH. Increased lipid peroxidation in sera of ALS patients: A potential biomarker of disease burden. Neurology. 2004;**62**(10):1758-1765

[88] Neel DV, Basu H, Gunner G, Chiu IM. Catching a killer: Mechanisms of programmed cell death and immune activation in amyotrophic lateral sclerosis. Immunological Reviews. 2022;**311**(1):130-150

[89] Fang X, Wang H, Han D, Xie E, Yang X, Wei J, et al. Ferroptosis as a target for protection against cardiomyopathy. Proceedings of the National Academy Science U S A. 2019;**116**(7):2672-2680

[90] Cadenas S. ROS and redox signaling in myocardial ischemia-reperfusion injury and cardioprotection. Free Radical Biology & Medicine. 2018;**117**:76-89

[91] Das DK, Engelman RM, Liu X, Maity S, Rousou JA, Flack J, et al. Oxygen-derived free radicals and hemolysis during open heart surgery. Molecular and Cellular Biochemistry. 1992;**111**(1-2):77-86

[92] Meerson FZ, Kagan VE, Kozlov YP, Belkina LM, Arkhipenko YV. The role of lipid peroxidation in pathogenesis of ischemic damage and the antioxidant protection of the heart. Basic Research in Cardiology. 1982;**77**(5):465-485

[93] Gao M, Monian P, Quadri N, Ramasamy R, Jiang X. Glutaminolysis and transferrin regulate ferroptosis. Molecular Cell. 2015;**59**(2):298-308

[94] Linkermann A, Skouta R, Himmerkus N, Mulay SR, Dewitz C, De Zen F, et al. Synchronized renal tubular cell death involves ferroptosis. Proceedings of the National Academy Science U S A. 2014;**111**(47):16836-16841

[95] Friedmann Angeli JP, Schneider M, Proneth B, Tyurina YY, Tyurin VA, Hammond VJ, et al. Inactivation of the ferroptosis regulator Gpx4 triggers acute renal failure in mice. Nature Cell Biology. 2014;**16**(12):1180-1191

[96] Lu L, Wu W, Yan J, Li X, Yu H, Yu X. Adriamycin-induced autophagic cardiomyocyte death plays a pathogenic role in a rat model of heart failure. International Journal of Cardiology. 2009;**134**(1):82-90

[97] Takemura G, Kanoh M, Minatoguchi S, Fujiwara H. Cardiomyocyte apoptosis in the failing heart--a critical review from definition and classification of cell death. International Journal of Cardiology. 2013;**167**(6):2373-2386

[98] Zhang T, Zhang Y, Cui M, Jin L, Wang Y, Lv F, et al. CaMKII is a RIP3 substrate mediating ischemia- and oxidative stress-induced myocardial necroptosis. Nature Medicine. 2016;**22**(2):175-182

[99] Li W, Feng G, Gauthier JM, Lokshina I, Higashikubo R, Evans S, et al. Ferroptotic cell death and TLR4/Trif signaling initiate neutrophil recruitment after heart transplantation. The Journal of Clinical Investigation. 2019;**129**(6):2293-2304

[100] Guo X, Zhou D, An P, Wu Q, Wang H, Wu A, et al. Associations between serum hepcidin, ferritin and hb concentrations and type 2 diabetes risks in a han Chinese population. The British Journal of Nutrition. 2013;**110**(12):2180-2185

[101] Altamura S, Mudder K, Schlotterer A, Fleming T, Heidenreich E, Qiu R, et al. Iron aggravates hepatic insulin resistance in the absence of inflammation in a novel db/db mouse model with iron overload. Molecular Metabolism. 2021;**51**:101235

[102] Wang X, Chen X, Zhou W, Men H, Bao T, Sun Y, et al. Ferroptosis is essential for diabetic cardiomyopathy and is prevented by sulforaphane via AMPK/NRF2 pathways. Acta Pharmaceutica Sinica B. 2022;**12**(2):708-722

[103] Zang H, Wu W, Qi L, Tan W, Nagarkatti P, Nagarkatti M, et al. Autophagy inhibition enables Nrf2 to exaggerate the progression of diabetic cardiomyopathy in mice. Diabetes. 2020;**69**(12):2720-2734. DOI: 10.2337/db19-1176

[104] Du S, Shi H, Xiong L, Wang P, Shi Y. Canagliflozin mitigates ferroptosis and improves myocardial oxidative stress in mice with diabetic cardiomyopathy. Frontiers in Endocrinology. 2022. DOI: 10.3389/fendo.2022.1011669

[105] Tuo QZ, Lei P, Jackman KA, Li XL, Xiong H, Li XL, et al. Tau-mediated iron export prevents ferroptotic damage after ischemic stroke. Molecular Psychiatry. 2017;**22**(11):1520-1530

[106] Li Q, Han X, Lan X, Gao Y, Wan J, Durham F, et al. Inhibition of neuronal ferroptosis protects hemorrhagic brain. JCI Insight. 2017;**2**(7):e90777

[107] Dietrich RB, Bradley WG Jr. Iron accumulation in the basal ganglia following severe ischemic-anoxic insults in children. Radiology. 1988;**168**(1):203-206

[108] Lipscomb DC, Gorman LG, Traystman RJ, Hurn PD. Low molecular weight iron in cerebral ischemic acidosis in vivo. Stroke. 1998;**29**(2):487-492

[109] Ding H, Yan CZ, Shi H, Zhao YS, Chang SY, Yu P, et al. Hepcidin is involved in iron regulation in the ischemic brain. PLoS One. 2011;**6**(9):e25324

[110] Park UJ, Lee YA, Won SM, Lee JH, Kang SH, Springer JE, et al. Blood-derived iron mediates free radical production and neuronal death in the hippocampal CA1 area following transient forebrain ischemia in rat. Acta Neuropathologica. 2011;**121**(4):459-473

[111] Patt A, Horesh IR, Berger EM, Harken AH, Repine JE. Iron depletion or chelation reduces ischemia/reperfusion-induced edema in gerbil brains. Journal of Pediatric Surgery. 1990;**25**(2):224-227

[112] Davis S, Helfaer MA, Traystman RJ, Hurn PD. Parallel antioxidant and antiexcitotoxic therapy improves outcome after incomplete global cerebral ischemia in dogs. Stroke. 1997;**28**(1):198-204

[113] Prass K, Ruscher K, Karsch M, Isaev N, Megow D, Priller J, et al. Desferrioxamine induces delayed tolerance against cerebral ischemia in vivo and in vitro. Journal of Cerebral Blood Flow and Metabolism. 2002;**22**(5):520-525

[114] Hanson LR, Roeytenberg A, Martinez PM, Coppes VG, Sweet DC, Rao RJ, et al. Intranasal deferoxamine provides increased brain exposure and significant protection in rat ischemic stroke. The Journal of Pharmacology and Experimental Therapeutics. 2009;**330**(3):679-686

[115] Speer RE, Karuppagounder SS, Basso M, Sleiman SF, Kumar A, Brand D, et al. Hypoxia-inducible factor prolyl hydroxylases as targets for neuroprotection by “antioxidant” metal chelators: From ferroptosis to stroke. Free Radical Biology Medical. 2013;**62**:26-36

[116] Chang CF, Cho S, Wang J. (−)-epicatechin protects hemorrhagic brain via synergistic Nrf2 pathways. Annals of Clinical Translational Neurology. 2014;**1**(4):258-271

[117] Rodríguez C, Sobrino T, Agulla J, Bobo-Jiménez V, Ramos-Araque ME, Duarte JJ, et al. Neovascularization and functional recovery after intracerebral hemorrhage is conditioned by the Tp53 Arg72Pro single-nucleotide polymorphism. Cell Death Differentiation. Jan 2017;**24**(1):144-154. DOI: 10.1038/cdd.2016.109. Epub 2016 Oct 21. PMID: 27768124; PMCID: PMC5260494

[118] Zille M, Karuppagounder SS, Chen Y, Gough PJ, Bertin J, Finger J, et al. Neuronal death after hemorrhagic stroke in vitro and in vivo shares features of ferroptosis and necroptosis. Stroke. 2017;**48**(4):1033-1043

[119] Zhang Z, Wu Y, Yuan S, Zhang P, Zhang J, Li H, et al. Glutathione peroxidase 4 participates in secondary

brain injury through mediating ferroptosis in a rat model of intracerebral hemorrhage. Brain Research. 2018;**1701**:112-125

[120] Handa P, Thomas S, Morgan-Stevenson V, Maliken BD, Gochanour E, Boukhar S, et al. Iron alters macrophage polarization status and leads to steatohepatitis and fibrogenesis. Journal of Leukocyte Biology. 2019;**105**(5):1015-1026

[121] Woodhoo A, Iruarrizaga-Lejarreta M, Beraza N, García-Rodríguez JL, Embade N, Fernández-Ramos D, et al. Human antigen R contributes to hepatic stellate cell activation and liver fibrosis. Hepatology. 2012;**56**(5):1870-1882

[122] Zhang Z, Yao Z, Wang L, Ding H, Shao J, Chen A, et al. Activation of ferritinophagy is required for the RNA-binding protein ELAVL1/HuR to regulate ferroptosis in hepatic stellate cells. Autophagy. 2018;**14**(12):2083-2103

[123] Atarashi M, Izawa T, Kuwamura M, Yamate J. The role of iron overload in the progression of nonalcoholic steatohepatitis. Nihon Yakurigaku Zasshi. 2019;**154**(2):61-65

[124] Tsurusaki S, Tsuchiya Y, Koumura T, Nakasone M, Sakamoto T, Matsuoka M, et al. Hepatic ferroptosis plays an important role as the trigger for initiating inflammation in nonalcoholic steatohepatitis. Cell Death & Disease. 2019;**10**(6):449

[125] Wang W, Green M, Choi JE, Gijón M, Kennedy PD, Johnson JK, et al. CD8+ T cells regulate tumour ferroptosis during cancer immunotherapy. Nature. 2019;**569**(7755):270-274

[126] Fan X, Li A, Yan Z, Geng X, Lian L, Lv H, et al. From Iron metabolism to ferroptosis: Pathologic changes in coronary heart disease. Oxidative Medicine and Cellular Longevity. 2022;**10**(2022):6291889

[127] Li Q, Zhao Z, Zhou X, Yan Y, Shi L, Chen J, et al. Ferroptosis: The potential target in heart failure with preserved ejection fraction. Cell. 2022;**11**(18):2842

[128] Zheng DW, Lei Q, Zhu JY, Fan JX, Li CX, Li C, et al. Switching apoptosis to ferroptosis: Metal-organic network for high-efficiency anticancer therapy. Nano Letters. 2017;**17**(1):284-291

[129] You L, Wang J, Liu T, Zhang Y, Han X, Wang T, et al. Targeted brain delivery of rabies virus glycoprotein 29-modified deferoxamine-loaded nanoparticles reverses functional deficits in parkinsonian mice. ACS Nano. 2018;**12**(5):4123-4139

Section 3

Autophagy

Chapter 6

Impaired Autophagy and Exosomes Release by Long-Term mTOR Pathway Activation Promotes Hepatocellular Carcinoma Occurrence and Invasion

Qirong Wen, Qingfa Zeng and Ting Li

Abstract

Mammalian target of rapamycin (mTOR) is highly expressed in various types of hepatocellular carcinoma (HCC). Clinically, HCC cases without inflammation and cirrhosis are also increasingly common, especially in patients with nonalcoholic fatty liver disease, more and more patients develop HCC, which is only characterized by hepatic steatosis. However, the molecular mechanisms underlying the development of non-inflammatory HCC remain unclearly. Our previous study demonstrated that overactivation of mTOR pathway in the liver promotes de novo lipid synthesis and eventually spontaneous formation of non-inflammatory HCC. The continuous activation of mTOR pathway, on the one hand, promotes the de novo synthesis of lipids, resulting in the production of a large amount of lipid in the liver; on the other hand, it inhibits autophagy, resulting in the inability of lipid to be removed in time and accumulate in the liver. Accumulated lipid peroxidation eventually develops into HCC. In addition, the continuously activated mTOR pathway inhibited the release of exosomes by reducing the expression of Rab27A, and *in vitro* experiments confirmed that hepatoma cells after Rab27A knockout were more prone to invasion and metastasis. The reduced release of exosomes may impair intercellular communication, especially with immune cells, thereby making HCC more prone to invasion and metastasis with less inflammation.

Keywords: mTOR pathway, liver cancer, autophagy, exosomes, mouse model

1. Introduction

Cancer is a type of disease that seriously threatens human health. Among all cancer types, liver cancer is a very common gastrointestinal malignancy and one of the leading causes of cancer-related deaths [1–3]. Despite impressive advances in medicine over the past few decades, due to the occult nature of liver cancer, the uncertainty of its pathogenesis, the lack of effective treatments [4, 5], early diagnosis

IntechOpen

of liver cancer, the survival rate, and prognosis are extremely poor, and the 5-year average survival rate is less than 10% [2, 3].

Rapamycin signaling pathway is abnormally up-regulated in about 50% of liver cancer patients [6]. The mTOR signaling pathway can not only regulate the metabolism of nutrients, such as nucleotides, lipids, and proteins but also inhibit autophagy and stimulate cell growth [7, 8]. Since the mTOR pathway plays a major role in a variety of metabolic and physiological processes, it also contributes to diseases and pathological conditions, such as aging, metabolic syndrome, and cancer when it is dysregulated [8]. As the global incidence of metabolic syndrome and obesity continues to rise, the accompanying liver cirrhosis and liver cancer are also increasing year by year, seriously threatening human life and health [9]. At the same time, there is also research evidence that lipid de novo synthesis plays a key role in the occurrence and development of human liver cancer, and data reveal that the mTOR pathway is the main regulatory pathway for abnormal lipid synthesis in liver cancer [10].

Our previous research report also found that chronic overactivation of mTOR pathway promotes de novo lipid synthesis in mice and eventually develops into hepatocellular carcinoma (HCC) [11]. Interestingly, the development of HCC in such mice was not accompanied by overt necroinflammation and liver fibrosis [12]. However, the pathogenesis of liver cancer is very complex, and the regulation of mTOR pathway is also very extensive. The involvement of mTOR pathway in regulating the occurrence and development of liver cancer may not be a single event but may involve the participation of multiple factors. Therefore, in this chapter, we will further explore the related mechanism of mTOR pathway involved in regulating the occurrence and development of liver cancer.

2. The mechanism of mTOR pathway involved in regulating the occurrence and development of liver cancer is very complex

2.1 Long-time chronic activation of mTOR pathway resulted in spontaneous HCC

2.1.1 HCC independent of long-term chronic injury and necroinflammation

The mTOR pathway is altered in various disease models and exhibits abnormal activation in tumor diseases, such as breast cancer, prostate cancer, lung cancer, liver cancer, kidney cancer, and lymphoma [13]. The TSC1/TSC2 heterodimer, consisting of tuberous sclerosis complex 1 (TSC1) and tuberous sclerosis complex 2 (TSC2), is an upstream inhibitor of mTOR pathway. Our previous data suggested that liver-specific knockout of TSC1 in mice leads to persistent activation of mTOR pathway that resulted in spontaneously HCC (**Figure 1A** and **B**). Histopathological features showed that the tumor cells were large with large nuclei and various staining, which was the pathological nuclear feature of malignant tumors. And only the cancer type of HCC was detected in these mouse models, indicating that the tumor was of hepatocyte origin and not of hepatic progenitor or bile duct epithelial origin (**Figure 1B**). Interestingly, these HCC model mice did not develop obvious inflammation and fibrosis, and the serum inflammatory factor levels did not increase significantly (**Table 1**), indicating that the continuous activation of mTOR pathway caused spontaneous HCC that was independent of long-term chronic injury and necroinflammation.

Studies have found that HCC is usually triggered by the death of liver cells, which usually leads to liver damage and secondary inflammatory response.

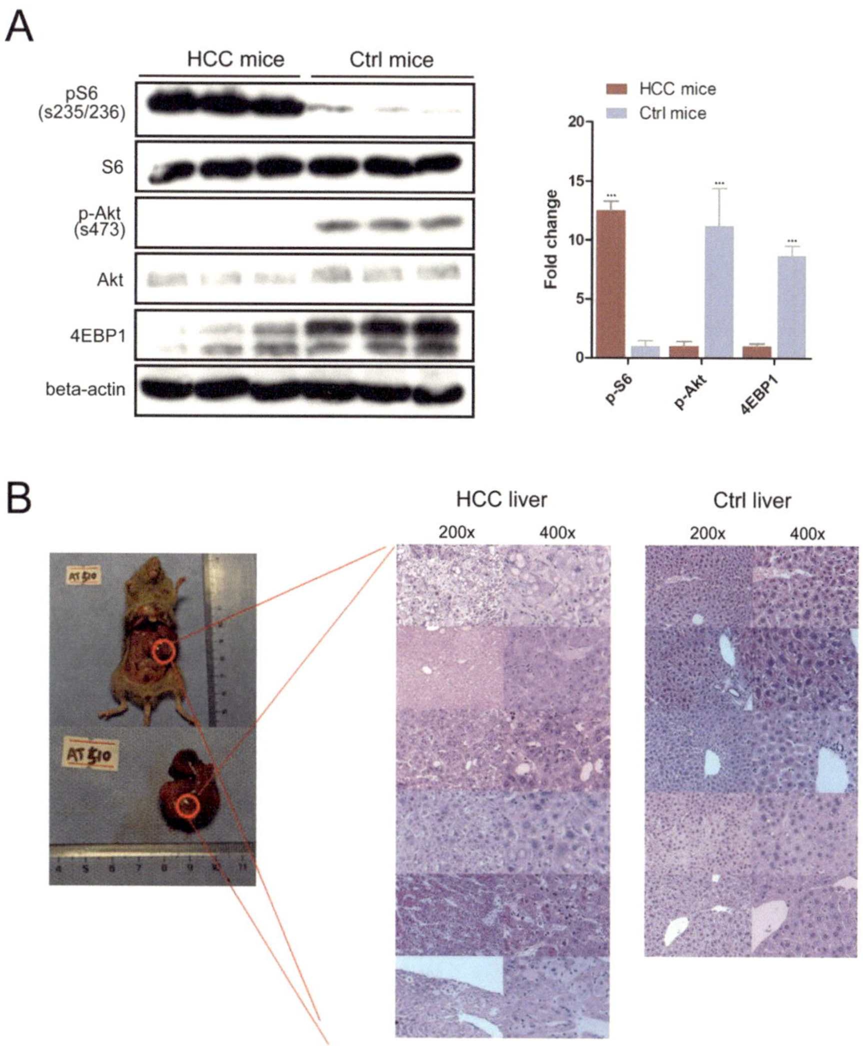

Figure 1.
*Persistent activation of mTOR pathway resulted in spontaneous HCC. (A) Western blots analysis showed the expression of mTOR pathway-related proteins in mouse liver tissue. Western blots were quantified based on at least 3 replicates. Error bars represent the SEM. ***P < 0.001. The resulting figure is cited from author's previously published article [11]. (B) the general picture of one spontaneous HCC mouse (left) and the pathological staining results of HCC mice and control mice (right), the magnifications are 200x and 400x.*

Therefore, majority of HCC tissues are accompanied by significant inflammation. Tumor inflammation is primarily caused by pro-tumor cytokines, including IL-6, that induces activation of the oncogenic transcription factor signal transducer and activator of transcription 3 (STAT3) in the hepatocytes, ultimately promoting compensatory proliferation of hepatocytes that have escaped cell death and subsequently promotes tumor development [14–16]. Activation of STAT3 further promotes IL-6 production and promotes inflammatory outbreaks. In a mouse model with liver-specific knockout of Raptor, a subunit of mTOR pathway, it was found that the increase of IL-6 and the activation of STAT3, ultimately promoted the occurrence and development of HCC, and showed mild liver cirrhosis in the process of developing

Factors	Ctrl mice (n = 5)	HCC mice (n = 9)	P-Value
IL-1a	52.87	49.76	0.6585
IL-1b	64.57	54.39	0.0904
IL-2	84.33	73.14	0.3589
IL-3	8.47	7.50	0.4328
IL-4	14.69	19.25	0.2418
IL-5	808.48	1083.47	0.4041
IL-6	319.35	283.91	0.5075
IL-9	8886.21	10211.62	0.1261
IL-10	32.61	31.45	0.7371
IL-12(p40)	9.21	13.04	0.2568
IL-12(p70)	128.93	146.81	0.2912
IL-13	3690.65	4507.50	0.2183
G-CSF	16.61	16.14	0.9514
GM-CSF	14.52	14.48	0.9923
IFN-g	5395.06	4226.12	0.1253
KC	186.65	189.51	0.9405
MCP-1	181.32	1260.69	0.2414
MIP-1a	42.22	50.83	0.6072
MIP-1b	367.00	340.20	0.2998
RANTES	838.56	810.17	0.7478
TNF-a	4568.78	3764.13	0.4491

Table 1.
The spectrum of inflammatory factors in the liver of mice.

into HCC [14]. In addition, STAT3 can also be activated by inhibiting signal transducer and activator of transcription 5 (STAT5), which promotes hepatocarcinogenesis through TGF-β [17, 18], and STAT5 also has a role in suppressing inflammation. However, no significant increase in inflammatory factors, including IL-6 and TNF-α was found in our mouse model. Meanwhile, accompanied by the activation of STAT5 phosphorylation and inhibition of STAT3 phosphorylation. Moreover, studies by others and our previous studies have demonstrated that mTOR pathway regulates phosphorylation of STAT3 and STAT5 [19]. Therefore, the spontaneous HCC model without necroinflammation might be developed through the activation of STAT5 with simultaneous inactivation of STAT3. Moreover, because there is no long-term necroinflammatory stimulus, the fibrosis of liver tissue is not very obvious.

2.1.2 The mTOR pathway regulates lipid de novo synthesis to mimic the progression of NAFLD to HCC

Since the first case of NAFLD-related HCC was reported in 1990 [20], there has been increasing evidence of an association between NAFLD and HCC, HCC is a common and lethal malignancy worldwide. Some risk factors (such as HBV, HCV, or alcoholism) are recognized in most cases of HCC, but there are also HCC that was not caused by these risk factors, and the incidence of such HCC is as high as 15–50%. In developed countries, these "unexplained" HCCs are mainly attributed to NAFLD [21]. Cancer cells are characterized by a shift in cellular metabolism to adapt to cancer cell growth and provide more energy, which is conducive to the continued development of carcinogenesis. The mTOR signaling pathway is also an important metabolic regulatory pathway. The mTOR pathway is abnormally activated in the liver and other tissues of patients with obesity, mTOR activation promotes fatty synthesis acid by upregulating the transcription, synthesis and nuclear transfer of sterol regulatory element-binding protein-1 (SREBP1) [22].

This spontaneous HCC mouse model also exhibits the features of metabolic disorders. We used the pathway enrichment analysis method on mouse liver tissue RNA-seq data to enrich the related enriched functional groups regulating the mTOR pathway. And found that metabolism-related pathways ranked first in all enriched pathways (**Figure 2**). These metabolic pathways mainly, include chemical carcinogenesis-related pathways, steroid hormone biosynthesis pathways, linoleic acid metabolism pathways, and cytochrome P450 (CYP450) metabolic pathways in xenobiotics.

In addition, the results metabolomic assay of liver tissue showed that oleic acid (C18:1) increased and stearic acid (C18:0) decreased, which increased membrane fluidity, promoting metabolism and proliferation. Stearic acid has hepatoprotective and anti-inflammatory potential and can reduce fibrosis after cholestasis-induced liver injury [23]. The anticancer effects of liver-specific Pten-deficient female mice are at least partly due to a marked reduction in the ratio of oleic to stearic acid in the liver [24]. Accumulation of fatty acids may interfere with cell signaling pathways and promote tumorigenesis by altering gene transcriptional regulation, which may be the mechanism by which NAFLD, characterized by fat accumulation, eventually develops into liver cancer.

In our mouse model, after treatment with rapamycin, the abnormal lipid metabolism was reversed, and the occurrence of HCC was also effectively reversed, further proved that abnormal lipid metabolism is involved in the occurrence and development of liver cancer.

2.2 Autophagy is involved in the regulation of the occurrence and development of liver cancer

2.2.1 Autophagy impairment in persistent mTOR pathway activation mouse model

In addition to regulating metabolism, the mTOR pathway is also a classic autophagy regulatory pathway. When the mTOR pathway is activated, it mediates the

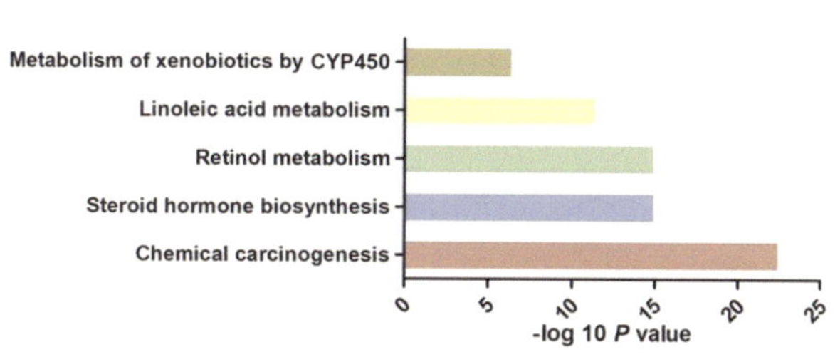

Figure 2.
Histographs of differentially expressed genes by pathway analysis.

phosphorylation of specific sites of ULK1 and Atg13 and inhibits the autophagy-promoting kinase activity of the ULK1 complex. When the mTOR pathway is inhibited, it is separated from ULK1, the phosphorylation of specific sites of ULK1 and Atg13 is released, the ULK1 complex is activated, and the activated ULK1 complex is then transferred to the isolation membrane of the endoplasmic reticulum, and autophagy is initiated [25].

Studies have found that AMPK/mTOR pathway-mediated autophagy activation is an important protective mechanism of NAFLD, mainly by inhibiting liver de novo adipogenesis, increasing fatty acid oxidation in the liver and promoting mitochondrial function/integrity in adipose tissue [26]. In the liver, autophagy, as a metabolic pathway, can regulate lipid accumulation in hepatic steatosis; while persistent mTOR pathway activation in hepatocytes leads to endoplasmic reticulum stress and autophagy defects, which are closely related to the occurrence and development of HCC. In our mouse model, sustained activation of hepatic mTOR pathway was found to lead to inhibition of autophagy. On the one hand, the continuous activation of mTOR pathway increases the de novo synthesis of lipids, and on the other hand, autophagy is inhibited, and the increased lipids cannot be efficiently metabolized and continuously accumulate in the liver. The study also believes that in the pathogenesis of NAFLD, autophagy is first activated and then inhibited, especially the continuous inhibition of autophagy will lead to the occurrence and further deterioration of NAFLD, which is consistent with our findings.

Furthermore, in the early stages of carcinogenesis, autophagy has significant cytoprotective and tumor suppressive potential. Dysfunction of this process is associated with an increased risk of cancer development. Therefore, we can also explain the phenomenon of spontaneous liver cancer in mice from the perspective of autophagy inhibition. The long-term continuous activation of the mTOR pathway keeps the autophagy in the liver of mice in a state of inhibition, and autophagy cannot effectively play the cytoprotective and tumor suppressive functions. Eventually, this leads to the occurrence of liver cancer. We demonstrated the above notion that rapamycin treatment in mice reversed autophagy impairment and prevented HCC development.

2.2.2 Activation of autophagy inhibits the invasion and metastasis of hepatoma cells in vitro

However, the role of autophagy in carcinogenesis has been controversial. Before tumorigenesis, effective autophagy can play cytoprotective and tumor suppressive effects, and autophagy can also be activated by tumor suppressor factors, such as PTEN, TSC, and DEPTOR [27, 28]. When tumors have already occurred, autophagy can further support tumor progression. Autophagy can promote tumor cell survival and malignant behavior by providing nutrients to tumor cells, thereby promoting tumor occurrence and development, while autophagy inhibition may make cancer cells. Sensitivity to metabolic stress conditions leads to tumor cell death [22, 29]. Many aggressive tumors require autophagy to facilitate important tumor-promoting processes [30].

However, some studies hold the different view. In a study by Zhang. et al., they found that SOCS5 promoted HCC cell migration and invasion *in vitro* by inactivating PI3K/Akt/mTOR pathway mediated autophagy. SOCS5 inhibition inhibited HCC cell migration and invasion *in vitro* by activating PI3K/Akt/mTOR pathway mediated autophagy. Dual inhibition of SOCS5 and mTOR further activates autophagy and exerts a more pronounced anti-metastatic effect in HCC cells [31].

In the spontaneous HCC mouse model, we found that on the basis of the continuous activation of the mTOR pathway and the inhibition of autophagy, some mice developed lung and intestinal metastasis of tumors (**Figure 3**). However, it was found that autophagy was activated and cell proliferation, invasion, and metastasis were decreased after treatment of liver cancer cells with rapamycin *in vitro* (**Figure 4A–D**). After the occurrence of tumors, autophagy may work with other mechanisms to maintain tumor homeostasis. When the effect of autophagy is relieved, it will destroy this homeostasis and promote tumor invasion and metastasis.

Future studies can pay more attention to the role of mTOR pathway mediated autophagy in different stages of liver cancer development, or explore other possible mechanisms to further clarify the mechanism of autophagy in different stages before and after tumorigenesis.

2.3 The occurrence and development of liver cancer involve complex mechanisms

The pathogenesis of tumors, including liver cancer, is very complex. In our spontaneous liver cancer mouse model, we found that in addition to abnormal STAT3/STAT5 pathway, lipid accumulation, and autophagy inhibition, there are many other abnormal manifestations.

2.3.1 Inhibition of exosomes secretion

It is well known that the mTOR pathway regulates autophagy to remove some cellular components, such as organelles to control cellular metabolism. The exosomes released and also transport part of the cell membrane and cellular components to the extracellular space, resulting in the loss of cellular contents. There is an overlap between autophagy and exosomes release, so the secretion of exosomes may also be regulated by the mTOR pathway. Our lab found that like autophagy, exosomes release is negatively regulated by the mTOR pathway in response to changes in nutrient and growth factor conditions, and the mTOR pathway mainly functions in the late stages of exosome biogenesis, possibly in the context of MVB (Multivesicularbody) and

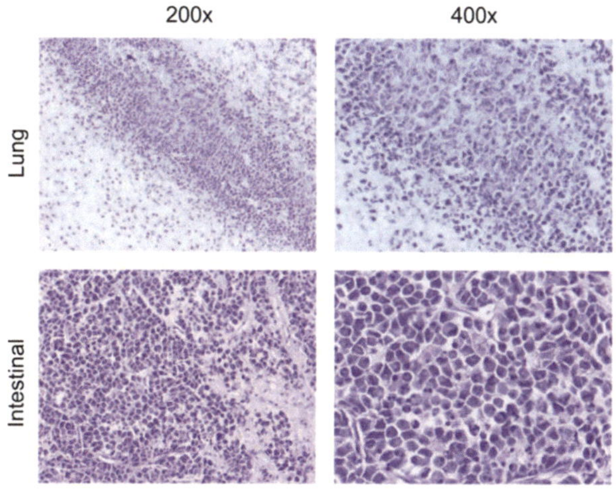

Figure 3.
The lung and intestinal metastasis in spontaneous HCC mouse. The magnifications are 200x and 400x.

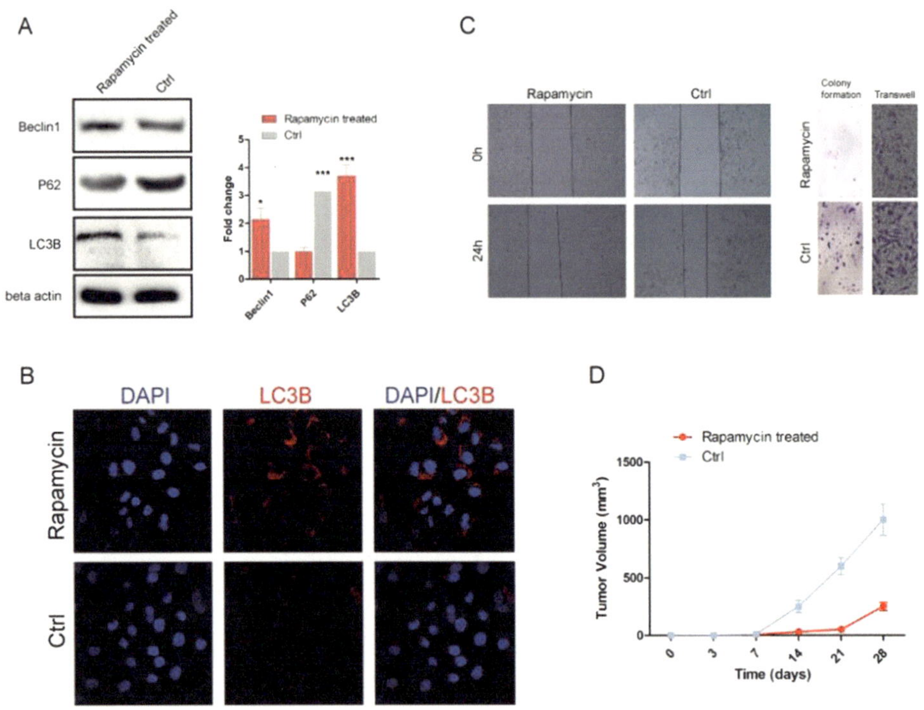

Figure 4.
*Autophagy activation inhibited tumor malignancy. (A) Western blots analysis showed the expression of autophagy-related proteins in Hep3B cells. Western blots were quantified based on at least 3 replicates. Error bars represent the SEM. ***P < 0.001. (B) Immunofluorescence analysis showed that rapamycin activates autophagy in Hep3B cells. (C) Wound healing experiments, clone formation, and transwell experiments confirmed that rapamycin treatment activates autophagy to inhibit the proliferation, invasion, and metastasis of Hep3B cells. (D) Tumor growth curves showed that autophagy activation inhibited tumor formation in nude mice.*

cytoplasmic membrane docking/fusion phase and validated in our mouse model, this inhibition was reversed after treatment with rapamycin [32]. We also detected a reduction of exosomes in the perisinusoidal space (Disse space) in this mouse model of spontaneous liver cancer in which knockout of TSC1 resulted in persistent mTOR pathway activation (**Figure 5A**). The inhibitory effect of mTOR pathway activation on the release of exosomes is mainly achieved by Rab27A, which does not change the content of exosomes. Rab27A and Rab27B are two small GTPases of the Rab family, which are involved in the docking and fusion of conditional MVB with the plasma membrane [33], which is crucial for the release of exosomes and the transmission of extracellular messengers, and it also regulates membrane homeostasis, lysosomal function, and autophagy.

The study found that cancer cells usually produce more exosomes than normal cells, exosomes from cancer cells have a strong ability to alter local and distant microenvironments, and exosomes from highly invasive tumor cells deal with low invasiveness tumor cells will enhance the latter's ability to invade and metastasize [34, 35]. In addition, tumor-derived exosomes can activate T cells, NK cells, or macrophages to activate immunity through direct or indirect antigen presentation [36]. Rab27A-overexpressing exosomes from tumor cells can further activate immunity, thereby promoting the proliferation of CD4+ T cells and exerting more effective anti-tumor immunity [37]. However, when Rab27A was silenced, Epithelial-mesenchymal

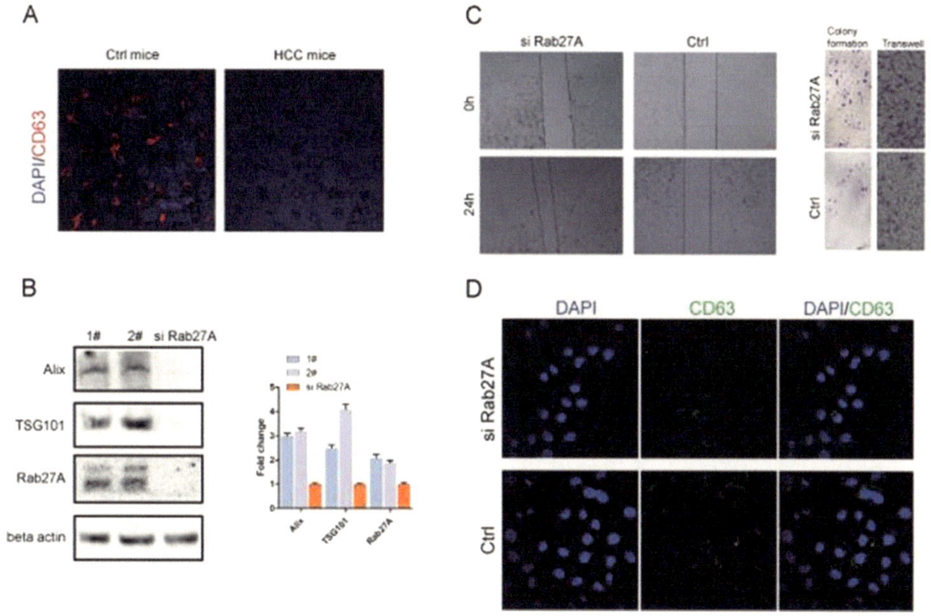

Figure 5.
Rab27A inhibited tumor malignancy. (A) Immunofluorescence analysis showed a reduction of exosomes in the Disse space in spontaneous HCC mouse model. (B) Western blots analysis showed the expression of Alix, TSG101, and Rab27A in 7721 cells. Western blots were quantified based on at least 3 replicates. (C) Wound healing experiments, clone formation, and transwell experiments confirmed that silencing Rab27A promoted the proliferation, invasion, and metastasis of 7721 cells. (D) Immunofluorescence analysis showed that silencing Rab27A inhibited the release of exosomes in 7721 cells.

transition (EMT) was induced through MAPK/ERK signaling pathway to promote cell migration, chemotaxis, and invasion, and the intrahepatic and lung metastasis increased [38]. Silencing the expression of Rab27A *in vitro* inhibited the release of exosomes and promoted the proliferation, migration, and invasion of tumor cells (**Figure 5B–D**). In our spontaneous liver cancer mouse model, the mTOR pathway was continuously activated, which continuously inhibits the release of exosomes, which may block intercellular communication from the perspective of exosomes, thereby failing to activate immunity and accelerate tumor development. The reason for the mild liver inflammation in this model mouse was also explained from the perspective of exosomes.

Recent studies have also shown the intersection of autophagy, exosome/amphoteric biogenesis, and exocytosis of extracellular vesicles [39, 40]. Autophagy is also involved in the exosome secretion process, changing not only the amount but also the content of exosomes. The effect of autophagy on the secretion of exosomes can be either promotion or inhibition, which may be closely related to the environment in which cells are located. This may partly explain the double-edged nature of autophagy in cancer progression.

2.3.2 Mitochondrial dysfunction

In 1956, Otto Warburg proposed that mitochondrial respiration defects were the potential basis of aerobic glycolysis and cancer [41], and the "Warburg effect" has been the basis of FDG-PET for tumor imaging. However, the mechanism of mitochondrial action in cancer was unclear, although mutations in the mitochondrial genome have been identified in human cancer specimens [42]. Mitophagy is a specialized autophagy that selectively degrades and eliminates excess or damaged mitochondria [43]. Oncocytomas are rare benign tumors of most epithelial cells characterized by massive accumulation of defective mitochondria due to pathogenic mtDNA mutations [44]. In addition, the TCA cycle is also one of the metabolic pathways that occur within mitochondria, and how mutations in mitochondrial TCA cycle enzymes lead to cancer by producing oncogenic metabolites [45, 46].

Mitochondrial dysfunction and alterations in the TCA cycle were also found in our spontaneous liver cancer mouse model. It was mainly reflected indirectly by the decreased expression of Ddit4 and Nupr1. Among them, Ddit4 reflects mitochondrial function, and when its expression is abnormal, it disrupts energy homeostasis and promotes tumorigenesis, while functionally, the activation of mTOR pathway and cell survival requires the inhibition of Ddit4, and the inhibition of Ddit4 contributes to the continuous activation of mTOR pathway and tumorigenesis cell survival. Nupr1 is a mitochondrial-deficient gene involved in regulating autophagy induced by lipotoxicity of excess fatty acid accumulation in cells [47]. Nupr1 has also been identified as a key regulator and metabolic switch in response to mitochondrial damage during liver cancer development [48]. Changes in TCA cycle were manifested in the increased levels of fatty acid metabolites involved and the decreased levels of glucose metabolites involved. These changes all work together to promote the swearing development of liver cancer.

2.3.3 FGF21 alteration is a late event in spontaneous HCC

We also found a decrease in fibroblast growth factor 21 (FGF21) protein levels in the tumor tissue of the spontaneous liver cancer model mice, but there was no difference in FGF21 between model mice and control mice before tumorigenesis, and the reversal of FGF21 protein levels was not evident after treatment with rapamycin. Perhaps the change of FGF21 is only a concomitant phenomenon after tumorigenesis; however, FGF21 has been confirmed to promote the occurrence and development of NAFLD.

FGF21 is mainly expressed in liver, thymus, adipose tissue, and pancreatic islet beta cells [49], and it was regulated by the PI3K/AKT pathway to reduce blood sugar, reduce liver fat deposition, and reduce body weight [50]. Studies have shown that when the body was in a state of fasting and starvation, the expression of FGF21 in the liver and adipose tissue was induced, resulted in an increase in liver lipolysis and hepatic glycogen synthesis; an increase in adipose tissue glucose uptake, a decrease in lipolysis and an increase in lipogenesis [51]. FGF21 also reduces hepatic triglyceride and cholesterol production by inhibiting SREBP2 [52]. FGF21 was not only involved in the regulation of hepatic fat metabolism but also negatively regulated by the mTOR signaling pathway [53]. In addition, FGF21 may also be involved in the regulation of liver cell polarity and affect the normal structure of liver tissue (**Figure 6**). FGF21 regulated bile acid metabolism and also inhibited the expression of CYP7A1, which was the first step in the conversion of cholesterol to bile acids. Bile synthesis and secretion can promote the formation of hepatocyte polarity. When bile acid metabolism was disordered, hepatocyte polarity was also cannot maintained, and dysbiosis of bile acid metabolism was an important indicator for the pathological diagnosis of NAFLD. Deletion of FGF21 can affect both bile acid metabolism and lipid metabolism and ultimately promote the occurrence and development of NAFLD.

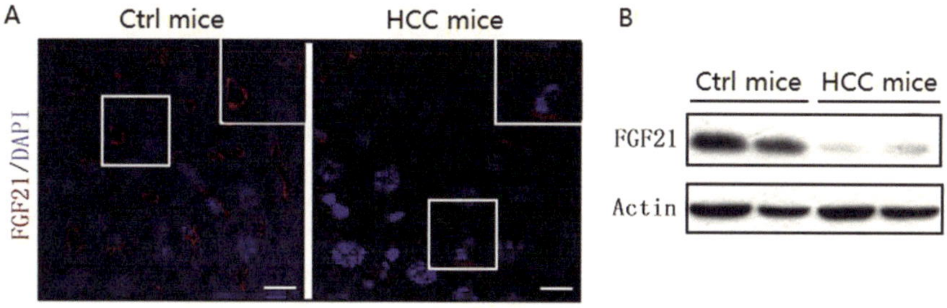

Figure 6.
Rab27A inhibited tumor malignancy. (A) Immunofluorescence analysis showed a reduction of FGF21 expression in spontaneous HCC mouse model. (B) Western blots analysis showed a reduction of FGF21 expression in spontaneous HCC mouse model.

In addition, FGF21 can aggravate tumor progression. Studies have found that reduced FGF21 protein levels are associated with cancerous hyperproliferation and abnormal p53, TGF-β/Smad signaling pathways during HCC development [54]; FGF21 can reduce hepatic fat accumulation and hepatocyte damage, while at the same time inhibiting inflammation and fibrosis and plays an important role in limiting progression of liver pathology from NAFLD/NASH to HCC [55, 56].

2.3.4 Ectopic expression of SLC22A7

Studies have found that SLC22A7 is highly expressed in liver cancer cells [57] and, SLC22A7 is associated with multicenter recurrence after liver cancer surgery, which may be achieved by regulating mitochondrial and oxidoreductase activities [58, 59]. In our spontaneous liver cancer mouse model, we found that SLC22A7 was not significantly altered at the protein level but was ectopically expressed. In normal liver, it was mainly expressed in the cell membrane of hepatocytes and in tumor liver, it was expressed in the cytoplasm (**Figure** 7).

2.3.5 Shows similar manifestations to chemical carcinogens

Further analysis of the results of transcriptome sequencing revealed that the spontaneous liver cancer mouse model has abnormal expression of many enzymes related genes, which are mainly involved in the detoxification process, including cytochrome P450 (CYP450) family, carboxylesterase (Ces), sulfotransferases (Sults), and UDP-glucuronyltransferases (Ugts). In the CYP450 family members, Cyp1a2, Cyp2b9, Cyp2c50, Cyp2c54, Cyp2c67, Cyp2e1, and Cyp3a16 expression decreased, while Cyp2b10 expression increased. CYP450 family-related enzymes are the key enzymes necessary for the phase I metabolism of exogenous substances in the liver. The Cyp450 family, such as Cyp1a2 and Cyp2e1 was a key enzyme in tumor transformation and mediates the metabolic activation of many carcinogens, which degrade xenobiotics, steroids, and fatty acids. Previous studies have found that the dysregulation of Cyp1a2 and Cyp2B9 mainly occurs in the liver of chemically carcinogenic mice [60, 61]. Another report found that Cyp2c50, Cyp2c54, and Cyp2c67 were significantly increased in the liver in a mouse model of chemically induced hepatocellular carcinoma [62]. Some researchers used Cyp2e1 knockout mice to study the effect of chemical carcinogenesis and found that Cyp2e1 knockout mice showed lower tumor incidence and diversity, indicating that the gene is a tumor protection related gene [63, 64].

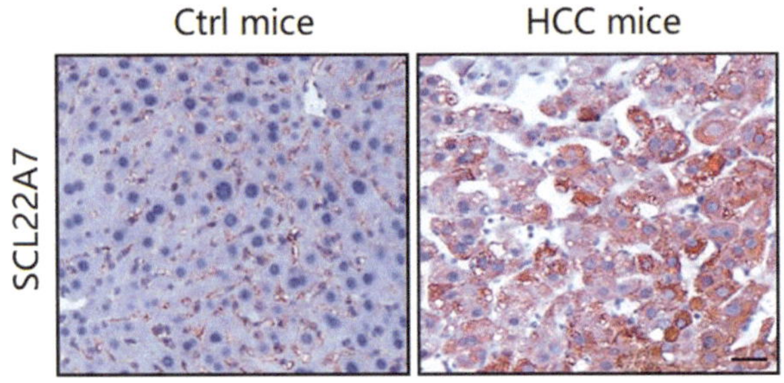

Figure 7.
Ectopic expression of SLC22A7. Immunohistochemical staining showed that SLC22A7 was mainly expressed in the cell membrane of hepatocytes, while wrong expressed to the cytoplasm in tumor cells of spontaneous HCC mouse model.

The first stage of tumorigenesis following mTOR pathway activation was analogous to chemical carcinogenesis, providing the primitive cells that both responsive to metabolic alterations and had a greater proliferative advantage over surrounding normal cells. Ultimately, these cells can be clonally expanded and transformed into cancer cells for immortal proliferation.

3. Conclusion

This chapter discusses some factors related to the mTOR pathway that may be involved in the occurrence and prognosis of HCC. There are multiple responsible changes in this spontaneous liver cancer model mouse, including accumulation of lipids, non-necrotizing inflammation, mitochondrial dysfunction, autophagy inhibition, exosome release inhibition, and protein ectopic expression. These changes may play a role before the occurrence of tumors, at the initial stage of tumors, or after the occurrence of tumors, and jointly promote the occurrence and development of liver cancer. Of course, the occurrence and development of HCC is a very complex process, and its related mechanism has always been a research hotspot, and new possible mechanisms will be reported over time.

Continued research is still needed to overcome difficult problems and ultimately be used for clinical treatment of liver cancer. Prevent the occurrence of liver cancer and reduce the recurrence of liver cancer after surgery. May the world be free from any cancer.

Author details

Qirong Wen[1], Qingfa Zeng[2] and Ting Li[3]*

1 Department of Gynecology and Obstetrics, Guangzhou Women and Children's Medical Center, Guangzhou Medical University, Guangzhou, China

2 R & D Department of Guangzhou All-perfect Biological Technology Co., Ltd., Guangzhou, China

3 Department of Hepatobiliary Surgery II, Zhujiang Hospital, Southern Medical University, Guangzhou, China

*Address all correspondence to: 373663456@qq.com

IntechOpen

© 2022 The Author(s). Licensee IntechOpen. This chapter is distributed under the terms of the Creative Commons Attribution License (http://creativecommons.org/licenses/by/3.0), which permits unrestricted use, distribution, and reproduction in any medium, provided the original work is properly cited.

References

[1] Sung H, Ferlay J, Siegel RL, Laversanne M, Soerjomataram I, Jemal A, et al. Global Cancer statistics 2020: GLOBOCAN estimates of incidence and mortality worldwide for 36 cancers in 185 countries. CA: A Cancer Journal for Clinicians. 2021;**71**(3):209-249

[2] Konyn P, Ahmed A, Kim D. Current epidemiology in hepatocellular carcinoma. Expert Review of Gastroenterology & Hepatology. 2021;**15**(11):1295-1307

[3] Bray F, Ferlay J, Soerjomataram I, Siegel RL, Torre LA, Jemal A. Global cancer statistics 2018: GLOBOCAN estimates of incidence and mortality worldwide for 36 cancers in 185 countries. CA: A Cancer Journal for Clinicians. 2018;**68**(6):394-424

[4] Forner A, Reig M, Bruix J. Hepatocellular carcinoma. Lancet. 2018;**391**(10127):1301-1314

[5] Yang JD, Hainaut P, Gores GJ, Amadou A, Plymoth A, Roberts LR. A global view of hepatocellular carcinoma: Trends, risk, prevention and management. Nature Reviews. Gastroenterology & Hepatology. 2019;**16**(10):589-604

[6] Sun EJ, Wankell M, Palamuthusingam P, McFarlane C, Hebbard L. Targeting the PI3K/Akt/mTOR pathway in hepatocellular carcinoma. Biomedicine. 2021;**9**(11)

[7] Yin Y, Hua H, Li M, Liu S, Kong Q, Shao T, et al. mTORC2 promotes type I insulin-like growth factor receptor and insulin receptor activation through the tyrosine kinase activity of mTOR. Cell Research. 2016;**26**(1):46-65

[8] Saxton RA, Sabatini DM. mTOR Signaling in growth, metabolism, and disease. Cell. 2017;**168**(6):960-976

[9] Ratziu V, Giral P, Jacqueminet S, Charlotte F, Hartemann-Heurtier A, Serfaty L, et al. Rosiglitazone for nonalcoholic steatohepatitis: One-year results of the randomized placebo-controlled fatty liver improvement with rosiglitazone therapy (FLIRT) trial. Gastroenterology. 2008;**135**(1):100-110

[10] Guertin DA, Stevens DM, Thoreen CC, Burds AA, Kalaany NY, Moffat J, et al. Ablation in mice of the mTORC components raptor, rictor, or mLST8 reveals that mTORC2 is required for signaling to Akt-FOXO and PKCalpha, but not S6K1. Developmental Cell. 2006;**11**(6):859-871

[11] Li T, Weng J, Zhang Y, Liang K, Fu G, Li Y, et al. mTOR direct crosstalk with STAT5 promotes de novo lipid synthesis and induces hepatocellular carcinoma. Cell Death & Disease. 2019;**10**(8):619

[12] Li T, Zhang G, Wang L, Li S, Xu X, Gao Y. Defects in mTORC1 network and mTORC1-STAT3 pathway crosstalk contributes to non-inflammatory hepatocellular carcinoma. Frontiers in Cell and Development Biology. 2020;**8**:225

[13] Okosun J, Wolfson RL, Wang J, Araf S, Wilkins L, Castellano BM, et al. Recurrent mTORC1-activating RRAGC mutations in follicular lymphoma. Nature Genetics. 2016;**48**(2):183-188

[14] Umemura A, Park EJ, Taniguchi K, Lee JH, Shalapour S, Valasek MA, et al. Liver damage, inflammation, and enhanced tumorigenesis after persistent

mTORC1 inhibition. Cell Metabolism. 2014;**20**(1):133-144

[15] Wang Q, Yu WN, Chen X, Peng XD, Jeon SM, Birnbaum MJ, et al. Spontaneous hepatocellular carcinoma after the combined deletion of Akt isoforms. Cancer Cell. 2016;**29**(4):523-535

[16] Galicia VA, He L, Dang H, Kanel G, Vendryes C, French BA, et al. Expansion of hepatic tumor progenitor cells in Pten-null mice requires liver injury and is reversed by loss of AKT2. Gastroenterology. 2010;**139**(6):2170-2182

[17] Mueller KM, Kornfeld JW, Friedbichler K, Blaas L, Egger G, Esterbauer H, et al. Impairment of hepatic growth hormone and glucocorticoid receptor signaling causes steatosis and hepatocellular carcinoma in mice. Hepatology. 2011;**54**(4):1398-1409

[18] Mueller KM, Themanns M, Friedbichler K, Kornfeld JW, Esterbauer H, Tuckermann JP, et al. Hepatic growth hormone and glucocorticoid receptor signaling in body growth, steatosis and metabolic liver cancer development. Molecular and Cellular Endocrinology. 2012;**361**(1-2):1-11

[19] Yokogami K, Wakisaka S, Avruch J, Reeves SA. Serine phosphorylation and maximal activation of STAT3 during CNTF signaling is mediated by the rapamycin target mTOR. Current Biology. 2000;**10**(1):47-50

[20] Powell EE, Cooksley WG, Hanson R, Searle J, Halliday JW, Powell LW. The natural history of nonalcoholic steatohepatitis: A follow-up study of forty-two patients for up to 21 years. Hepatology. 1990;**11**(1):74-80

[21] Starley BQ, Calcagno CJ, Harrison SA. Nonalcoholic fatty liver disease and hepatocellular carcinoma: A weighty connection. Hepatology. 2010;**51**(5):1820-1832

[22] Wang H, Liu Y, Wang D, Xu Y, Dong R, Yang Y, et al. The upstream pathway of mTOR-mediated autophagy in liver diseases. Cell. 2019;**8**(12)

[23] Pan PH, Lin SY, Ou YC, Chen WY, Chuang YH, Yen YJ, et al. Stearic acid attenuates cholestasis-induced liver injury. Biochemical and Biophysical Research Communications. 2010;**391**(3):1537-1542

[24] Anezaki Y, Ohshima S, Ishii H, Kinoshita N, Dohmen T, Kataoka E, et al. Sex difference in the liver of hepatocyte-specific Pten-deficient mice: A model of nonalcoholic steatohepatitis. Hepatology Research. 2009;**39**(6):609-618

[25] Kim J, Kundu M, Viollet B, Guan KL. AMPK and mTOR regulate autophagy through direct phosphorylation of Ulk1. Nature Cell Biology. 2011;**13**(2):132-141

[26] Zhong J, Gong W, Chen J, Qing Y, Wu S, Li H, et al. Micheliolide alleviates hepatic steatosis in db/db mice by inhibiting inflammation and promoting autophagy via PPAR-gamma-mediated NF-small ka, CyrillicB and AMPK/mTOR signaling. International Journal of Immunopharmacology. 2018;**59**:197-208

[27] Hou W, Liu J, Chen P, Wang H, Ye BC, Qiang F. Mutation analysis of key genes in RAS/RAF and PI3K/PTEN pathways in Chinese patients with hepatocellular carcinoma. Oncology Letters. 2014;**8**(3):1249-1254

[28] Chen X, Xiong X, Cui D, Yang F, Wei D, Li H, et al. DEPTOR is an in vivo tumor suppressor that inhibits prostate tumorigenesis via the inactivation

of mTORC1/2 signals. Oncogene. 2020;**39**(7):1557-1571

[29] Czaja MJ, Ding WX, Donohue TM Jr, Friedman SL, Kim JS, Komatsu M, et al. Functions of autophagy in normal and diseased liver. Autophagy. 2013;**9**(8):1131-1158

[30] Hao M, Yeo SK, Guan JL. Autophagy inhibition perturbs ERBB2 trafficking and abolishes tumorigenesis in ERBB2-driven breast cancer. Autophagy. 2021;**17**(4):1059-1060

[31] Zhang M, Liu S, Chua MS, Li H, Luo D, Wang S, et al. SOCS5 inhibition induces autophagy to impair metastasis in hepatocellular carcinoma cells via the PI3K/Akt/mTOR pathway. Cell Death & Disease. 2019;**10**(8):612

[32] Zou W, Lai M, Zhang Y, Zheng L, Xing Z, Li T, et al. Exosome release is regulated by mTORC1. Advanced Science (Weinh). 2019;**6**(3):1801313

[33] Kowal J, Tkach M, Thery C. Biogenesis and secretion of exosomes. Current Opinion in Cell Biology. 2014;**29**:116-125

[34] Demory Beckler M, Higginbotham JN, Franklin JL, Ham AJ, Halvey PJ, Imasuen IE, et al. Proteomic analysis of exosomes from mutant KRAS colon cancer cells identifies intercellular transfer of mutant KRAS. Molecular & Cellular Proteomics. 2013;**12**(2):343-355

[35] Yang Y, Li CW, Chan LC, Wei Y, Hsu JM, Xia W, et al. Exosomal PD-L1 harbors active defense function to suppress T cell killing of breast cancer cells and promote tumor growth. Cell Research. 2018;**28**(8):862-864

[36] Zhang HG, Zhuang X, Sun D, Liu Y, Xiang X, Grizzle WE. Exosomes and immune surveillance of neoplastic lesions: A review. Biotechnic & Histochemistry. 2012;**87**(3):161-168

[37] Li W, Mu D, Tian F, Hu Y, Jiang T, Han Y, et al. Exosomes derived from Rab27aoverexpressing tumor cells elicit efficient induction of antitumor immunity. Molecular Medicine Reports. 2013;**8**(6):1876-1882

[38] Chen L, Guo P, He Y, Chen Z, Chen L, Luo Y, et al. HCC-derived exosomes elicit HCC progression and recurrence by epithelial-mesenchymal transition through MAPK/ERK signalling pathway. Cell Death & Disease. 2018;**9**(5):513

[39] Deretic V, Jiang S, Dupont N. Autophagy intersections with conventional and unconventional secretion in tissue development, remodeling and inflammation. Trends in Cell Biology. 2012;**22**(8):397-406

[40] Salimi L, Akbari A, Jabbari N, Mojarad B, Vahhabi A, Szafert S, et al. Synergies in exosomes and autophagy pathways for cellular homeostasis and metastasis of tumor cells. Cell & Bioscience. 2020;**10**:64

[41] Warburg O. On respiratory impairment in cancer cells. Science. 1956;**124**(3215):269-270

[42] Stewart JB, Alaei-Mahabadi B, Sabarinathan R, Samuelsson T, Gorodkin J, Gustafsson CM, et al. Simultaneous DNA and RNA mapping of somatic mitochondrial mutations across diverse human cancers. PLoS Genetics. 2015;**11**(6):e1005333

[43] Randow F, Youle RJ. Self and nonself: How autophagy targets mitochondria and bacteria. Cell Host & Microbe. 2014;**15**(4):403-411

[44] Gasparre G, Romeo G, Rugolo M, Porcelli AM. Learning from oncocytic

tumors: Why choose inefficient mitochondria? Biochimica et Biophysica Acta. 2011;**1807**(6):633-642

[45] Lu C, Thompson CB. Metabolic regulation of epigenetics. Cell Metabolism. 2012;**16**(1):9-17

[46] Parker SJ, Metallo CM. Metabolic consequences of oncogenic IDH mutations. Pharmacology & Therapeutics. 2015;**152**:54-62

[47] Jia SN, Lin C, Chen DF, Li AQ, Dai L, Zhang L, et al. The transcription factor p8 regulates autophagy in response to palmitic acid stress via a mammalian target of rapamycin (mTOR)-independent Signaling pathway. The Journal of Biological Chemistry. 2016;**291**(9):4462-4472

[48] Lee YK, Jee BA, Kwon SM, Yoon YS, Xu WG, Wang HJ, et al. Identification of a mitochondrial defect gene signature reveals NUPR1 as a key regulator of liver cancer progression. Hepatology. 2015;**62**(4):1174-1189

[49] Huang Z, Xu A, Cheung BMY. The potential role of fibroblast growth factor 21 in lipid metabolism and hypertension. Current Hypertension Reports. 2017;**19**(4):28

[50] Yu D, Ye X, Wu Q, Li S, Yang Y, He J, et al. Insulin sensitizes FGF21 in glucose and lipid metabolisms via activating common AKT pathway. Endocrine. 2016;**52**(3):527-540

[51] Li Y, Wong K, Giles A, Jiang J, Lee JW, Adams AC, et al. Hepatic SIRT1 attenuates hepatic steatosis and controls energy balance in mice by inducing fibroblast growth factor 21. Gastroenterology. 2014;**146**(2):539-49 e7

[52] Li Q, Wang H, Zhang C, Tong R, Chen H, Qie R. Ethyl acetate extract of sappanwood alleviates experimental atherosclerosis in rats through changes in FGF21 and SREBP-2 expression. International Journal of Clinical and Experimental Pathology. 2020;**13**(2):220-229

[53] Estall JL, Ruas JL, Choi CS, Laznik D, Badman M, Maratos-Flier E, et al. PGC-1alpha negatively regulates hepatic FGF21 expression by modulating the heme/rev-Erb(alpha) axis. Proceedings of the National Academy of Sciences of the United States of America. 2009;**106**(52):22510-22515

[54] Zhang Q, Li Y, Liang T, Lu X, Liu X, Zhang C, et al. Loss of FGF21 in diabetic mouse during hepatocellular carcinogenetic transformation. American Journal of Cancer Research. 2015;**5**(5):1762-1774

[55] Singhal G, Kumar G, Chan S, Fisher FM, Ma Y, Vardeh HG, et al. Deficiency of fibroblast growth factor 21 (FGF21) promotes hepatocellular carcinoma (HCC) in mice on a long term obesogenic diet. Molecular Metabolism. 2018;**13**:56-66

[56] Tillman EJ, Rolph T. FGF21: An emerging therapeutic target for non-alcoholic steatohepatitis and related metabolic diseases. Frontiers in Endocrinology (Lausanne). 2020;**11**:601290

[57] Libra A, Fernetti C, Lorusso V, Visigalli M, Anelli PL, Staud F, et al. Molecular determinants in the transport of a bile acid-derived diagnostic agent in tumoral and nontumoral cell lines of human liver. The Journal of Pharmacology and Experimental Therapeutics. 2006;**319**(2):809-817

[58] Kudo A, Mogushi K, Takayama T, Matsumura S, Ban D, Irie T, et al. Mitochondrial metabolism in the

noncancerous liver determine the occurrence of hepatocellular carcinoma: A prospective study. Journal of Gastroenterology. 2014;**49**(3):502-510

[59] Yasui Y, Kudo A, Kurosaki M, Matsuda S, Muraoka M, Tamaki N, et al. Reduced organic anion transporter expression is a risk factor for hepatocellular carcinoma in chronic hepatitis C patients: A propensity score matching study. Oncology. 2014;**86**(1):53-62

[60] Muguruma M, Nishimura J, Jin M, Kashida Y, Moto M, Takahashi M, et al. Molecular pathological analysis for determining the possible mechanism of piperonyl butoxide-induced hepatocarcinogenesis in mice. Toxicology. 2006;**228**(2-3):178-187

[61] Waalkes MP, Liu J, Chen H, Xie Y, Achanzar WE, Zhou YS, et al. Estrogen signaling in livers of male mice with hepatocellular carcinoma induced by exposure to arsenic in utero. Journal of the National Cancer Institute. 2004;**96**(6):466-474

[62] Graves JP, Gruzdev A, Bradbury JA, DeGraff LM, Edin ML, Zeldin DC. Characterization of the tissue distribution of the mouse Cyp2c subfamily by quantitative PCR analysis. Drug Metabolism and Disposition. 2017;**45**(7):807-816

[63] Kang JS, Wanibuchi H, Morimura K, Gonzalez FJ, Fukushima S. Role of CYP2E1 in diethylnitrosamine-induced hepatocarcinogenesis in vivo. Cancer Research. 2007;**67**(23):11141-11146

[64] Villeneuve JP, Pichette V. Cytochrome P450 and liver diseases. Current Drug Metabolism. 2004;**5**(3):273-282